Tourist Mobility and Advanced Tracking Technologies

Routledge Advances in Tourism

EDITED BY STEPHEN PAGE, *University of Stirling, Scotland*

Tourist Mobility and Advanced Tracking Technologies

Noam Shoval and Michal Isaacson

Routledge
Taylor & Francis Group
New York   London

First published 2010
by Routledge
270 Madison Ave, New York, NY 10016

Simultaneously published in the UK
by Routledge
2 Park Square, Milton Park, Abingdon, Oxon OX14 4RN

Routledge is an imprint of the Taylor & Francis Group, an informa business

© 2010 Taylor & Francis

Typeset in Sabon by IBT Global.
Printed and bound in the United States of America on acid-free paper by IBT Global.

Library of Congress Cataloging in Publication Data
Shoval, Noam.
  Tourist mobility and advanced tracking technologies / by Noam Shoval and Michal Isaacson.
  p. cm.—(Routledge advances in tourism; 19)
  Includes bibliographical references and index.
  1. Tourism—Social aspects. 2. Tourism—Psychological aspects. 3. Travelers—Psychology. 4. Tracking radar. I. Isaacson, Michal. II. Title.
  G155.A1U53 2009
  910.72—dc22
                    2009017558

ISBN10: 0-415-96352-4 (hbk)
ISBN10: 0-203-86937-0 (ebk)

ISBN13: 978-0-415-96352-7 (hbk)
ISBN13: 978-0-203-86937-6 (ebk)

Contents

PART IV
Concluding Thoughts

Figures

Tables

Preface

In the summer of 2003, I had been awarded a research grant to investigate tourism in three small historic cities, the Old City of Akko being one of them. One of the main themes I was interested in was the spatial activity of tourists throughout these cities. Michal joined me as a research assistant and together we set out to examine patterns of spatial activity in the Old City of Akko. The task at hand was not simple; Old Akko is a small area made up of many tiny alleyways and passages with poor signage and poor maps. We initially considered using a standard time-activity diary, but realized very quickly that it would not be effective. After returning from a walk around the city one day, we found we had great difficulty plotting the route we had taken on a map. If two geographers, we realized, had difficulty recalling their route and recognizing it on a map, what chance was there that participants in a study would be able to fulfill the same task?

Next we contacted private investigators, reasoning those who track people professionally must have some insight as to the effective tracking of visitors to the Old City of Akko. The private investigators did not offer any real solutions to our problem. We realized that we would need to find a solution ourselves and arranged meetings with companies that marketed tracking services.

The first meeting was held with a marketing executive working for Ituran Ltd., an Israeli company that sells tracking services to car owners. The company, whose technology will be introduced later in the book, was very generous and willing to cooperate with us, allowing us to conduct several tests on their system. At the same time, we met with the Israeli representatives of well-known GPS manufacturers. They too let us test their equipment but were not sure how we would be able to modify the technology to make it suitable for distribution to tourists who could thus passively record their visits throughout the city. The companies were accustomed to clients such as armies and security forces in different part of the world, clients who could affix a GPS antenna to a soldier's helmet; that clearly would not work with tourists.

However, the results of our initial tests with the GPS receivers were very positive and we quickly understood that the study in the Old City of Akko would benefit greatly from the use of GPS receivers. In the end, the solution to our problems came from the Israeli branch of Motorola, which supplied us with nine GPS location kits with data-storing abilities that—despite their bulky appearance—did perform reasonably well and enabled us to carry out a successful project in the Old City of Akko.

From that point on, the road that has stretched before us has been full of adventures and new discoveries. Stumbling on using tracking technologies has brought our way many new research topics and ideas, in the realm of tourism and in other areas, that would not have been possible otherwise. We are lucky to have been involved in this emerging field of research right from the beginning. We sincerely hope that readers of this book will share our enthusiasm for the possibilities that these methods open up in the study of spatial activity in general and the study of the spatial activity of tourists in particular.

Acknowledgments

This book is the result of five years of ongoing research, and we are indebted to many people, companies, and organizations that supported this project. We will begin by thanking Michael Hall, who envisioned and initiated the very idea of writing this book in 2005; this book could not have been written without his vision and encouragement. We wish to acknowledge the support of The Hebrew University of Jerusalem and the Israel Science Foundation (grant No.832/03), which served as the catalyst for the implementation of tracking technologies in tourism research. We wish to thank Tamar Soffer for her help in drawing most of the figures in this book, Adi Bennun for his advice on GIS-related issues, Joshua Rosenbloom for his comments on different parts of the manuscript, and Avi Toltzis for his instruction and advice on legal issues in relation to Chapter 9. We also wish to thank Deena Glickman for her important help with the English linguistic and stylistic editing and Ben Holtzman, our editor from Routledge.

We must also thank our research partners in Catalonia and Hong Kong—Salvador Anton Clave and Paolo Russo of the School of Tourism and Leisure (EUTO), University Rovira i Virgili, and Bob McKercher of the School of Hotel and Tourism Management at the Hong Kong Polytechnic University—for their insight and productive collaboration in tourist-tracking projects and for letting us present some of the findings of studies conducted in collaboration in this book.

Noam Shoval spent a sabbatical year (2007–2008) at the Department of Geography of the Heidelberg University with Peter Meusburger and Tim Freytag as an Alexander von Humboldt Fellow. This enabled him to dedicate time to the writing of the book. Many thanks are due to the Humboldt Foundation for making this possible. Noam Shoval spent the second part of 2008 as a guest of the Catalan Government at the School of Tourism and Leisure of the University Rovira i Virgili with Salvador Anton Clave and Paolo Russo. This precious time, and especially the intellectual exchange with Paolo and Salvador, contributed a lot to the writing of this book.

We also wish to thank the following companies for allowing us to test various tracking devices and services: Ituran Ltd., Mirs Comunications Ltd., Paz Logistics Ltd., and Partner Communications Ltd.

On a less formal, but no less important, level we wish to thank Amit Birenboim, Tamar Edry, Kinneret Cohen, and Yuval Kantor, members of Noam Shoval's research team, for all of their support and essential input along the years in various tracking projects.

Finally, we wish to thank our spouses and children for their support and understanding while we were writing this book.

Noam Shoval and Michal Isaacson
Jerusalem, Israel
March 2009

1   Introduction

A decade ago, David Fennell concluded his article on tourists' spatial behavior in the Shetland Islands with the following note: In the future, he reflected, information about the behavior of tourists in time and space would be best "accomplished by adopting and modifying the 'radiotelemetry' technology used for many years in the natural sciences . . . for the purpose of tracking specific animal species in parks and protected area environments" (1996, 827–828). While exploring the behavior of tourists in time and space, Fennell, like others before him, encountered a range of problems; his observation about the best possible method for researching the subject was a result of the challenges he faced.

Of the various problems associated with the traditional methods employed in the gathering of information on tourists' spatial and temporal behavior, the most common are those relating to the level of accuracy and/or the validity of the data collected. As a result, despite the proliferation of research in tourism over the past few decades, and even though it is a fundamental feature of the tourism phenomenon, relatively little attention has been paid to the spatial and temporal behavior of tourists (Dietvorst 1995; Thornton et al. 1997; Shaw et al. 2000; Forer 2002; Shaw and Williams 2002).

This dearth of research is especially troubling given that it is widely recognized that the movement of tourists has profound implications for infrastructure and transport development, tourism product development, marketing strategies, the commercial viability of the tourism industry, and the management of the social, environmental, and cultural impacts of tourism. Past research has focused primarily on the movement of tourists between destinations or from source markets to destination areas, applying concepts of distance decay, market access, and the valuation of time. Methodological problems have prevented most researchers from undertaking similar studies of smaller areas, such as urban destinations.

Until recently, the most common method for gathering information on human time-space patterns was the time-space diary. This method provides a systematic record of the way in which individuals occupy their time in space over a limited period, be it a few hours, a day, or a week (Anderson 1971). While time-space diaries have been used to great effect (see, for example, Goodchild and Janelle 1984; Janelle et al. 1988) they do have several disadvantages as research tools. In particular, time-space diaries require that the subjects are actively involved in the process of data collection by recording, in detail and at length, their activities throughout the entire experiment (Thornton et al. 1997). Since participants often fail to record their actions faithfully, the data obtained are often of questionable credibility (Szalai 1972).

In recent years, the rapid development and availability of small, cheap, and reliable tracking devices has led to a growing volume of spatial research in general and in tourism studies in particular. These studies will be discussed later in this book; however, it should be noted that the efforts to develop commercial applications for tourists, including location-aware mobile information systems or location-aware electronic guidebooks, have been underway since the end of the 1990s (Schilling et al. 2005; Ten Hagen et al. 2005).

Global Positioning System (GPS) devices offer researchers the opportunity of continuous and intensive high-resolution data collection in time (seconds) and space (meters) for long periods of time; this was never possible before in spatial research. GPS and other tracking technologies are now used in a wide variety of fields aside from tourism, such as environmental health (Phillips et al. 2001; Elgethun et al. 2003); the medical field, in such subjects as Alzheimer's disease (Miskelly 2004; Miskelly 2005; Shoval et al. 2008), physiology (Terrier and Schutz 2005), and cardiology (Le Faucheur et al. 2008); and as a tool to assist in navigation for visually impaired and blind pedestrians (Golledge et al. 1991; Golledge et al. 1998; Maeda et al. 2002). However, to date, most of the research conducted based on material gathered by advanced technologies has been in the field of transportation studies, usually in regard to tracking the spatial paths of motor vehicles (see, for example, Zito et al. 1995; Quiroga and Bullock 1998; Wolf et al. 2001; Bohte and Maat 2009). The collection of data and the study of the spatial activity of pedestrians using advanced technologies have been less common.

One possible explanation for this state of affairs is that gathering data from pedestrians by this means is more complicated than doing so from motor vehicles. Whereas for a car the advanced tracking system is simply one more accessory which is easily installed and does not affect the nature of the data collected, in the case of pedestrians the tracking system must be both small and "passive," ensuring that it does not disrupt or affect the subject's normal behavior; these requirements are often difficult to meet. This, however, has now changed, thanks to the technological advance that has enabled the manufacturing of small, lightweight, highly sensitive GPS receivers.

Three reasons can be given for the rapid development of tracking technologies in recent decades. All three relate to processes that took place in the United States.

1) THE DEVELOPMENT OF GLOBAL POSITIONING SYSTEMS (GPS) FOR MILITARY PURPOSES

Although the Russian GPS (Glonass) is in operation and plans for a European GPS (Galileo) are underway, the best known and most commonly used Global Positioning System is that belonging to the U.S. Department of Defense (DOD). It was originally conceived as a military navigation system. Fully operational since 1994 (Kaplan 1996), the system was initially available to military personnel only, with the DOD deliberately downgrading the satellites' civilian signal in order to deny civilians access to its system. In May 2000 the DOD terminated the Selective Availability (SA) procedure, as it was known, opening up the system for individuals and commercial applications across the globe. The result: Usage of the American GPS became so widespread that the term GPS is, at present, virtually synonymous with the DOD system.

2) THE INVENTION OF THE CELLULAR PHONE AND ITS RAPID DISSEMINATION WITHIN THE PRIVATE SECTOR

At the same time that GPS was being developed by the public sector in the United States, the private sector finished establishing infrastructure for the operation of cellular phones. Although the commercial use of cellular communications commenced at the beginning of the 1980s, use was limited primarily to business purposes due to the high price of both the service and the devices. Cellular phone prices began to drop drastically in the mid-1990s and today, in the developed world, cell phones are owned by people of all ages, professions, and income levels. Cell phone penetration in the developed world recently crossed the 80 percent mark (Eurostat 2005). In 2005, the United Kingdom had a 106 percent penetration rate, second only to Israel with a penetration rate of 112 percent (World Bank 2006). It is interesting that in the United States, where the "cellular revolution" began, the penetration rates is only around the 70 percent level. In recent years, the penetration of this form of communication technology has accelerated in many parts of the developing world as well. It is expected that by 2010 more than 50 percent of the world's population will own a cellular phone, at which time we will be able to refer to human society as a whole as a Cellular Society (Shoval 2007).

Operating a cellular phone network requires that the network operator can constantly detect the subscriber's proximity to a specific transceiver ("cell"). This enables the operator to transmit incoming and outgoing calls to and from the user's handset. To this end, the cell phone regularly communicates with the transceivers in its immediate vicinity, even when no calls are being made. This feature allows for the tracking of the device.

3) THE ENHANCED 911 SYSTEM AND THE IMPLEMENTATION OF NEW FCC REGULATIONS

In 1996, the Federal Communications Commission introduced a program designed to improve the 911 emergency service provided to mobile phone users. The program, which, when fully operational, will allow 911 dispatchers to identify a caller's geographic location automatically, was launched in two phases. During Phase 1, which concluded in 1998, service carriers were required to report to the FCC both the caller's mobile telephone number and the location of the transceiver that received the call signal. This narrowed down the caller's position to an area a mere few square kilometers in size (i.e., the radius around the transceivers). Phase 2, which has yet to be fully implemented, obligates the mobile phone companies to pinpoint the caller's location to within 50 to 100 meters.

E911, as the program is known, has already sparked a growing market in tracking technologies. There has been a marked rise in the development of tracking technologies, the aim being to produce cost-effective technologies that fall within the FCC's guidelines (Foroohar 2003).

Interest in utilizing location technology to assist in the work of emergency services is growing. The steps taken by the FCC have been followed by the European Union's "Coordination Group on Access to Location Information for Emergency Services" (C.G.A.L.I.E.S). The group issued a 2001 proposal for the E112 program, with objectives similar to the American program (Ludden et al. 2002).

The ability to collect time-space data at such high resolutions in time and space for long periods of time opens up the possibility of drawing new lines of inquiry and creates opportunities to formulate new research questions that could not be asked previously. Could the potential impact of tracking technologies in the spatial sciences be compared one day to the impact of the introduction of high-resolution digital platforms in other fields, such as the MRI in medicine, the electron microscope in chemistry and biology, or the Ikonos earth observation satellite in remote sensing? Maybe this "prophecy" is a truly wild exaggeration, maybe not.

This book is the first book to be written about the implementation of advanced tracking technologies for the research of tourists' outdoor movements in time-space and their activities and probably the second one on this topic in the social sciences. The edited volume *Urbanism on Track* was probably the first such contribution in the social sciences (Schaick and Spek 2008); it focused on the potential contribution of tracking technologies for urban studies.

This book will describe, discuss, and evaluate the new technologically based methodologies for tracking pedestrians and motor vehicles in the context of tourism research, planning, and management.

STRUCTURE OF THE BOOK

The book is comprised of four main parts. Part I focuses on theoretical and methodological issues in tourist spatial behavior. Part II describes the principal relevant tracking technologies. Part III, which is the most important and also the most extensive part in the book, deals with the application of tracking technologies to the research of tourist mobility. Part IV elaborates on the issues of privacy and ethics in relation to tracking people and draws conclusions, summarizing the book's main findings.

Part I Theoretical and Methodological Issues of Tourists' Spatial Behavior

Chapter 2 reviews the literature relevant for understanding tourist spatial behavior. After an introduction on the growth of tourism in recent decades and on the centrality of tourists' time-space activities for understanding the tourist phenomenon, it continues with a review of Torsten Hägerstrand's concept of time geography, which has immense relevance for understanding the temporal and spatial patterns of tourists in destinations. The chapter introduces a conceptual model that aims that combines the different variables into one framework explaining the factors that influence tourist time-space activity in a destination. The differences between individual and organized tourism are then discussed as the most clear-cut division between tourist types.

Chapter 3 presents the primary methods traditionally employed in data collection of tourist time-space activities, including direct observation and non-observational techniques. Naturally, various "time-space budget" techniques are presented and discussed in greater detail as the main data-collection methods in the pre-tracking-technology days. The second part of the chapter focuses on data visualization methods of tourists' time-space activities. The starting point is Hägerstrand's famous time-space aquarium, which makes possible the presentation of space and time in a single diagram. Following this discussion, two types of pattern aggregation methods are presented: quantitative pattern aggregation and visual pattern aggregation. The chapter concludes with the presentation of several time-space models.

Part II Available Tracking Technologies

Chapter 4 describes the development of different land-based tracking technologies during the second half of the twentieth century. Land-based tracking systems are local tracking systems, featuring a series of antenna stations, also known as radio frequency (RF) detectors, distributed throughout a specific area. The chapter also presents the Cell Sector Identification (CSI) method in detail. This is the technology used to identify the location of a particular cellular phone within a cellular network. Given the prevalence of cellular networks and the ubiquity of cellular phones, this technology becomes an ideal tool for collecting data on tourists' spatial activity. Other technologies which are presented in detail are the angle of arrival (AOA) method and the time difference of arrival method (TDOA).

Chapter 5 explains the basic features of the various satellite navigation systems. It first reviews the systems available today and scrutinizes the technology on which they are based.

Part III Application of Tracking Technologies to Research on Tourist Mobility

This part, which includes three chapters, is the core of the book. Chapter 6 underlines the challenges and demands involved in applying the tracking technologies discussed in the previous two chapters to the study of tourist mobility. The first part of the chapter presents the results of several experiments that took place on a variety of geographical scales, in which tourists were tracked using different types of tracking technologies. The results of the experiments lead the authors to a determination of which method is the most suitable for each of the different geographical scales. The second part of the chapter matches the proper technology and equipment to the relevant temporal and spatial scale and resolves various questions regarding the sampling of the tourists.

Chapter 7 introduces the opportunities tracking technologies offer in deepening the understanding we have regarding the spatial behavior of tourists within a destination. The data presented were all collected using GPS, primarily because it was found to be the most suitable for time-space research of tourists at the time of writing, as explained in Chapter 6. The chapter begins by describing what can be learned from analyzing data collected from one tourist. It then discusses the possibilities for "real-time analysis" and the integration of spatial data into tourist questionnaires and interviews. The chapter continues with an explanation of the manipulation of spatial data to create different variables that can be analyzed by regular statistical methods. The following section of the chapter discusses typologies of tourists—typologies that are created using spatial data by applying a sequence alignment method that originates in biochemistry and typologies that are created without using spatial data but that can be enriched by adding the understanding of the spatial activity of these types of tourists.

Chapter 8 presents the contribution of aggregative data obtained from GPS receivers and cellular phones in order to better understand the impact of visitors on destinations. Such analysis could facilitate decisions such as where to set up new attractions and where to promote private-sector tourist services. In addition, this type of analysis could assist in creating tourism management policies to reduce congestion in hitherto overcrowded and over-exploited sites and to generally enhance the destinations' physical and social carrying capacity.

Section IV Concluding Thoughts

Tracking people raises moral and ethical questions, specifically in regard to the way in which devices may impinge upon individuals' right to privacy. Chapter 9 is devoted to elaborating on the moral and ethical issues related to the research. This is not a new issue, as even today commercial mobile phone companies can identify the location of their cell phone users.

Chapter 10 concludes the book, summarizing the main issues that were covered and discussing possible future research agendas.

Appendix

The appendix of this book explains some of the technical and practical issues related to the analysis of data obtained by advanced tracking technologies. Specifically, the chapter explains how obtained data are integrated into geographic information systems (GIS). Close attention is paid to the data's precision; how to use data drawn from different geographic projections; the accuracy of maps thus obtained; and the problem of factoring in the passage of time when using GIS data.

Part I

Theoretical and Methodological Issues of Tourists' Spatial Behavior

2   Theoretical Aspects of Tourists' Spatial Behavior

Research into human spatial behavior has flourished since the 1970s, with researchers conducting extensive empirical studies, which, in turn, yielded considerable theoretical advances. Their findings, most notably those of Hägerstrand (1970), Anderson (1971), Chapin (1974), Parks and Thrift (1980), and Hanson and Hanson (1981), were eventually published jointly with other subsequent research in the field in book form (Golledge and Stimson 1987; Golledge and Timmermans 1988; Golledge and Stimson 1997).

Over the past twenty years, there has been a marked rise in the number of studies devoted to analyzing place and space, a subject that has become an area of key interest within the social sciences (Goodchild et al. 2000; Kwan 2002a). This development owes much to recent advances in spatial technologies, most notably that of geographic information systems (GISs; Kwan et al. 2003).

This chapter contains an overview of the existing research in the field of tourists' spatial behavior. The first part of the chapter covers the growth of tourism and its impact on destinations, visitor mobility, and tourist practices; the data analysis methods employed in the study of tourists' time-space activities; and the potential contribution of the approach of time geography to tourism research. The second part of the chapter focuses on the existing information in academic literature about the impact of various variables on the spatial activity of tourists. The third part of the chapter focuses on the difference in activity between the individual tourist and organized tourism; this variable has an immense impact on the differences in time-space activities of tourists.

THE GROWTH OF TOURISM AND ITS IMPACT ON DESTINATIONS

Several concurrent economic, social, and technological processes resulted in the fact that tourism grew sharply in the second half of the twentieth century (Shachar 1995). Tourism was transformed from the exclusive luxury of the elite social classes it had been for centuries (Towner 1996) into a widespread phenomenon constituting part of the lifestyle of practically everyone in the developed world. This change in the very nature of tourism is reflected in a dramatic growth in international flow of travel, whereas in 1950 a total of slightly over 25 million tourists crossed international borders, this number exceeded 900 million in 2007.

Almost forty years ago, geographer and city planner Sir Peter Hall offered an excellent description of the rising importance of tourism in the economies of cities and in their urban planning, claiming that the "age of mass tourism is the biggest single factor for change in the great capitals of Europe, and in many smaller historical cities too, in the last 30 years of this century" (Hall 1970, 445). Indeed, the increasing tourist flows have served to irrevocably alter many locations. Numerous airports, for instance, were transformed from mere landing strips with small terminals into massive complexes that include shopping malls, hi-tech industrial parks, and hotels (Gottdiener 2000). Huge mega-resorts began to emerge, such as Las Vegas and Orlando in the United tates, the Gold Coast in Australia, Cancún in Mexico, and the Costa del Sol in Spain (Mullins 1991; Gladstone 1998). Capital cities and global financial centers have registered enormous growth, notably in business-oriented travel (Braun 1992; Law 1996). Similarly, historical cities have become magnets for tourism to such an extent that their physical and social carrying capacities are actually placed in jeopardy (Canestrelli and Costa 1991; Borg et al. 1996; Ashworth and Tunbridge 2000; Russo 2001; Page and Hall 2003). These issues have become factors in the broader context of visitor mobility in urban areas.

VISITOR MOBILITY AND TOURIST PRACTICES

Sightseeing, walking, shopping, and sitting in restaurants and cafés are widely recognized as the major activities of urban tourists. Although these activities appear to be clearly set in time and space, so far relatively little attention has been paid to visitor mobility within the fields of human geography and tourism research (Dietvorst 1995, 163; Thornton et al. 1997; Shaw et al. 2000).

Besides the fact that tourism research is a relatively new field of study, the dearth of research on this subject can be attributed to the methodological complexity involved in studies of this kind (Pearce 1988 and 2001; Meng et al. 2005). Firstly, it is difficult to locate the tourist when he or she enters and leaves a city or region due to the absence of defined entry and exit points. Secondly, the term "tourist" includes a wide variety of different types of tourists that are distinguishable from one another by their interests, the purpose of their visits, and their time budgets, among other factors, so that in order to illustrate the tourist's spatial behavior, the different types of tourists in the city must first be identified. Thirdly, the funding requirements for such surveys have restricted the wide implementation of empirical research on tourists' time-space activities (Forer 2002, 24).

Studies that do address spatial-temporal visitor activities are usually rather descriptive and mostly conducted at case study level; they rarely attempt to deal with the factors which form the basis of tourist mobility. Two recent exceptions are the important contributions of Lew and McKercher (2006) and McKercher and Lau (2008). Certain studies however, have focused on specific features with regard to their impact on spatial activity: for example, religion (Shachar and Shoval 1999), the purpose of the visit (Montanari and Muscara 1995), gender (Scraton and Watson 1998; Carr 1999), the number of visits to the city (Oppermann 1997), or the organization of the tourists (as individuals or a group; Chadefaud 1981). Nevertheless, a greater theoretical framework that serves to explain the interplay of several underlying factors for spatial activity of tourists in an urban setting has yet to be developed.

A better understanding of the logic of visitor activities in time and space could not only serve a number of practical purposes in tourism industries, planning, and management, but also develop the existing concept of time geography (Hägerstrand 1953; 1970) and considerably enlarge the theoretical foundations of tourism research. However, the traditional methods of collecting data on tourist spatial behavior have proved to be problematic—most obviously with regard to their accuracy and validity (discussed in detail in the next chapter).

DATA ANALYSIS METHODS EMPLOYED IN THE STUDY OF TOURISTS' TIME-SPACE ACTIVITIES

Though the aforementioned studies all make use of time-space data for the examination of tourists and visitors in their destinations, they differ in their research objectives and types of data analysis. It is possible to divide them into the following categories:

1. Descriptive analysis of tourists' movement and time allocation: In this type of analysis, the temporal and spatial patterns of tourists—either individual tourists or, more often, groups of tourists—are presented. This analysis is hardly ever the stand-alone objective of a study. It is usually just a preliminary phase of a more sophisticated analysis. Ferrell (1996), for example, presented in his paper the time-space allocation of two groups of visitors: general interest and special interest groups to the Shetland Islands.
2. Detection of explanatory and predictive factors for tourists' temporal and spatial behavior patterns: These studies aim to uncover factors that can explain the spatial visitation patterns of tourists. More conventional statistical analyses such as cluster analysis and correlations

are usually used in these studies. Kemperman et al. (2004), who looked for the influence of repeat visits in a theme park, provide an example of such a study.

3. Creation of typologies: The goal of such studies is to distinguish groups of tourists based on their temporal and spatial patterns of behavior. One of the most famous works of typology creation in tourism is Cohen's (1972) work that defines general types of tourists: organized mass tourists, individual mass tourists, explorers, and drifters. Cohen's paper is theoretical and certainly not based on any time-space data, but it does constitute a milestone for this type of research in tourism. A study that does make use of time-space data in order to distinguish between different types of tourists' activities is Dietvorst's (1994) research, which defines three different types of tourists according to their temporal and spatial patterns of activity.

4. Understanding tourists' decision-making and choices: These studies endeavor to understand the motives and factors that influence visitors' decision-making and choices. The time-space data in this context are used to determine the actual behavior of the tourists. Possible motives and causes that are related to the tourists' choices can be socioeconomic, demographic, psycho-cognitive, or related to various constraints. Though at times studies in this category use similar analytical methods to those of the explanatory and predictive factors researches (see category 2 above), they have a different theoretical framing. Thornton et al. (1997), for example, showed that the presence of young children in a group influences the decisions of a group, which adjusts its activities in order to accommodate the needs of the child.

5. Spatial cognition / abilities exploration: This type of research aims to improve knowledge regarding human spatial abilities such as navigation, orientation, and perception. An example from the field of tourism research can be found in Xia et al. (2008) who suggested four possible models of wayfinding strategies. Examining the proposed models on visitors to the Koala Conservation Centre on Phillip Island (Australia) using GPS devices, one of the main goals of the research was to use the findings to assist in the provision of wayfinding aids.

6. Movement patterns and flow: This type of work strives to identify and understand repeat movement patterns of tourists within a specific location. McKercher and Lau (2008), for example, discovered seventy-eight discrete movement patterns of tourists in Hong Kong. They then divided these patterns into eleven movement styles based on territoriality factor (length traveled from the hotel) and trip intensity (number of stops made during the trip).

7. Destination consumption: Hot spots, congestion, and the temporal and spatial impact of tourists' behavior on sites and recreational destinations are of interest for this group of studies. The tourist's behavior in these studies is not directly observed; rather, the reflection of

his or her activity through the spatial consumption of the location is investigated. Shoval (2008) used this approach to explore the impact of the visitors on the Old City of Akko.

It is important to note that some studies can be classified in several of these categories simultaneously.

TIME GEOGRAPHY AND ITS POTENTIAL CONTRIBUTION TO TOURISM RESEARCH

Time geography, focusing on the constraints and trade-offs that occur when people find themselves having to divide a limited amount of time between various activities in space, is one of the earliest analytical perspectives used to examine patterns of human activity (Miller 2005, 17). This is an important cornerstone for the conceptualization and visualization of tourist flows in space and time, as will be described later in this chapter and this book. Time geography was developed by the Swedish geographer Torsten Hägerstrand, who developed the basic tenets in the 1960s, 1970s, and early 1980s, together with his associates at the University of Lund, known collectively as the Lund School (Gregory 2000, 830). Beyond Sweden's borders, researchers such as Allan Pred, Nigel Thrift, and Anthony Giddens helped with the international diffusion of time-geographic thought. In particular, with his structuration theory and thoughts on space-time, Giddens made time geography known to a wider circle of researchers (Lenntorp 1999, 57). As a result, analysis of human activities in space-time burgeoned, not only among geographers but among transport researchers as well (Kwan 2004).

During the 1990s, however, interest in the field, at least among geographers, gradually faded, while in transport studies there were a number of "activity-based analysis" projects that explicitly drew upon time-geographic notions (e.g., Kondo and Kitamura 1987; Kitamura et al. 1990). However, the past decade has seen a resurgence in time-geographic studies. The spread of increasingly sophisticated geographic information systems (GIS), capable of providing detailed computational representations and more precise measurements of basic time-geographic entities including space-time paths and prisms, persuaded a growing number of researchers to return to the time geography fold (Miller 1991; Kwan 1998, 1999a, 1999b, 2000; Timmermans et al. 2002; Kwan and Lee 2004; and Miller 2005). The development of new digital information technologies, such as cellular phones, wireless Personal Digital Assistants (PDAs), Location-based Services (LBS), Global Positioning System (GPS) receivers, and radiolocation methods, multiplied the volume and improved the spatial-temporal resolutions of the empirical data to a degree previously unimagined, thus aiding the recent revival of the field (Kwan 2000; Miller 2003; Miller 2005; Kwan 2004; Raubal et al. 2004; and Shoval and Isaacson, 2006).

Torsten Hägerstrand's death in 2004 also propelled time geography up the geographical community's agenda, with, for example, the publication of a special volume of Geografiska Annaler (86B, 4) dedicated to him and an assessment of his scientific contributions.

In tourism research, the application of Hägerstrand's theoretical and pragmatic framework for visualization and analysis of time-space activities of tourists was done only scarcely despite its relevance to the field. Dietvorst's (1994; 1995) works on tourism to historic cities, Forer's (2002) implementation regarding flows of visitors to New Zealand, and especially Hall's (2005) conceptual work on the application of Hägerstrand's thought to tourism research are good examples for the future potential of this framework for tourism studies.

It is possible to use this approach in order to explain a tourist's spatial activity, though many of the factors existing for the tourist differ from those of the individual in his or her natural surroundings. In addition, in contrast with Hägerstrand's model, the tourist has much choice and a lot of free time, and therefore the tourist activity expresses more of his or her cultural background and personality and less of the "traditional" factors as understood by Hägerstrand. In place of these factors there are other constraints that result from the length of the stay or the primary purpose of the visit. It would not be surprising if we found that an entire range of socioeconomic factors were of no importance whatsoever for the explanation of a tourist's spatial activity. It is indeed possible to relate to constraints such as income, gender, and age as factors that are influential on the person in his or her natural surroundings; however, upon leaving the natural surroundings for a short time, the person may be freed from the limitations these factors place on him or her.

THE IMPACT OF VARIOUS VARIABLES ON THE SPATIAL ACTIVITY OF TOURISTS

The variables that are generally considered to be important in relation to the spatial activity of individual tourists can be divided into two main groups: The first group of factors relates to the character of the specific trip (length of visit, individual tourist or part of a group, etc.) and the second group is associated with the specific characteristics of the tourists (Jefferson and Lickorish 1988; Vanhove 1989).

Variables Related to the Character of the Trip

Purpose of visit: The central parameter for segmenting tourists into types, employed by national and international agencies and in academic research, is the "main purpose of the visit" (Dietvorst 1994; Montanari and Muscara 1995; Page and Hall 2003). Purpose of visit has a direct impact on the spectrum of possibilities available to the tourist: tourists who travel for business or to visit friends and relatives will be less likely to visit tourist sites than tourists who travel for the specific purpose of touring and sightseeing.

Length of stay: Despite the widespread use of this variable in the tourism industry and in various statistical sources, the literature on the spatial activity of tourists in cities has yet to deal with this variable. By projection from the time-space approach in geography (Hägerstrand 1970), it could be hypothesized that the amount of time that a tourist spends in the city will have a significant impact on his or her "spectrum of activities."

Number of visit to the destination: Murphy and Oppermann (1997) suggested that the more often a tourist visits a particular city, the fewer the visits to tourist sites on each visit (given the same duration of stay). Clearly, once familiar with the main tourist sites, the tourist will be less interested in visiting the same sites again and again (for a contrary finding, see Richardson and Crompton 1988).

Organization of the trip: The spatial activity of individual tourists and the geographic range of their activities in a city will be completely different from that of organized groups, as they are personally responsible for selecting the particular tourist sites to be visited (see, for example, Chadefaud 1981 on tourism to Lourdes).

Variables Related to the Tourist

Country of origin: Despite the widespread use of this variable, its efficiency in differentiating between tourist types is questionable. Some researchers view this variable as problematic because, as a result of increased migration to Western societies, the major source of most international tourism, tourists are becoming increasingly more heterogeneous. Consequently, this variable is becoming less and less relevant (Veal 1992; Dann 1993). On the other hand, research conducted on specific groups has shown differences in ethnic background to have an impact on the behavior and spatial activity of tourists at the tourist destination (Moore 1985; Richardson and Crompton 1988; Jansen-Verbecke 1991; Jules-Rosette 1994; Pizam and Sussman 1995; Pizam and Jeong 1996).

Gender: Even in the developed world there is a significant difference between the spatial activity of men and women in daily life (see Knox 1994, 283–290, for a review of this topic). These differences are the result of the unofficial constraints placed on women due to their more active role in chores related to taking care of household and children. Regarding leisure, several studies have shown that a difference between genders exists (Scraton and Watson 1998; Hall and Page 1999). This finding is hardly surprising, because these leisure activities take place within the constraints of daily life. These same considerations have been used in relation to tourism in explaining decisions pertaining to holiday destinations and activity at the destination itself (Bell 1991; Adler and Brenner 1992; Hughes 1997; Hughes 1998; Frew and Shaw 1999; Shaw and Williams 2002).

Age structure of the travel group: Cooper (1981) found that it is possible to differentiate the spatial activity of small groups of individual tourists according to age.

Religion: In destinations with religious attractions, tourists (and pilgrims) from different religions will have differential activity patterns. Research conducted in Jerusalem, which is an important center for the three main monotheistic religions, has demonstrated differences in spatial activity among tourists (and pilgrims) on the basis of religion (Shachar and Shoval 1999) as well as subdivisions within the same religion (for instance, between Catholics and Protestants; Bowman 1991).

Income: The income variable plays an important role in the social sciences in general; geography is no exception. In tourism research, the income level of tourists is likely to affect their spatial activity (Cooper 1981).

Education: In contrast to the situation regarding permanent place of residence, there have been no empirical findings regarding the relationship between education and the spatial activity of tourists. In studies of patterns of consumption in general, it was found that people who are more educated and affluent will show a greater tendency to attend cultural events or visit museums and heritage sites (see, for example Bourdieu and Darbel 1991, who studied the characteristics of visitors to European art museums, and Prentice 1993 and Light and Prentice 1994, who found a high correlation between affluence and visits to heritage sites in England and Wales). However, all of these studies concentrated on the activity of the local population and not the tourist population.

Personality type: Plog's (1973; 1987) classification of tourists by personality type was originally proposed to explain how personality structure is likely to influence the choice of travel destination. Plog's typology found support in Debbage's (1991) study of the spatial activity of tourists in a resort in one of the Bahamian islands.

External Variables

Weather: Variables related to weather conditions during a visit, such as a heat wave, a snowstorm, or a tropical storm (not to mention more extreme examples), can have a significant impact on the time-space activity of a tourist in a destination.

Transportation: Another type of variable relates to the nature of the transport infrastructure or the layout of the tourist sites in the city or region.

Crowding: The level of crowding in various sites may have an impact on the time-space activity of the tourist. For example, a long line at the entrance to a museum could change the plans for the day, so the museum will be visited later that day, on another day, or even not at all.

Recent work by the French geographer Rémy Knafou and his group "Mobilités, Itinéraires, Tourismes" (MIT) started to conceptualize tourism as a system that is produced through relations between tourists, tourist practices, and tourist places (Knafou and Stock 2003). German geographer Tim Freytag, a member of Knafou's group, graphically presented the various variables and constraints on the time-space activity of tourists in a destination during their visit (see Figure 2.1).

Empirical Findings Regarding the Importance of Different Variables

Nearly a decade ago, a study of 1,638 individual tourists' time-space activities in the cities of Jerusalem and Tel Aviv (Israel) enabled the identification of factors that bear an impact on the spatial activity of individual tourists (Shoval 2001; 2002). The most significant characteristics explaining spatial activity were found to be quite similar in both cases despite the very different characteristics of the two destinations. As expected, the most influential variables are those related to the character of the trip, such as length of stay in the city, main purpose of visit, and number of visit. Contrary to expectation, socioeconomic variables (gender, age, income, education, and country of origin) were found to be of lesser (or no) importance. It should be noted that in Jerusalem, as expected, religion played a major role in the consumption of tourist sites.

The "country of origin" variable was found to be of no importance in explaining the spatial activity of tourists in Jerusalem and Tel Aviv. This finding is not surprising due to the ever-increasing heterogeneity of this variable in tourism-generating countries in Europe and North America, as discussed earlier. Nor was "level of income" found to be an important factor in explaining the spatial activity of tourists. While this factor may influence the decision to visit a city, or which tourist services to consume (hotels and restaurants), while in the city it does not influence the tourist's spatial activity. Contrary to expectation, "education," which was expected to bear some influence—at least regarding visits to museums—was found to be of no importance either in Jerusalem or in Tel Aviv. This finding raises interesting issues regarding the assumption that education, quantified by "years of study," reflects potential interest in cultural events and visits.

Yet another interesting result concerns the finding that "gender" was of no importance in explaining the spatial activity of tourists in either of the two cities. This most probably derives from the fact that during a visit to the city as a tourist, many of the constraints prevailing in the individual's daily life in his or her regular place of residence cease to exist. It might even be claimed that tourism actually "liberates" women from chores associated with the household and children. The difference between men and women was found to exist for only one type of site in Tel Aviv, the shopping and

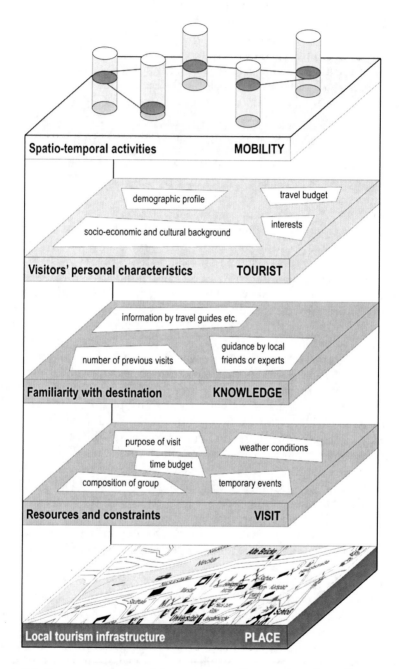

Figure 2.1 Conceptual model of the resources and constraints that affect tourists' spatial activity.

leisure sites. The "age" variable was also found to be relevant to the shopping and leisure sites, in this case in Jerusalem as well as Tel Aviv. This finding was expected in view of the different patterns of entertainment consumption in different stages of the lifecycle. Religious affiliation was found to be an important factor in explaining the spatial activity of tourists in Jerusalem. This finding was expected in view of the fact that the city is an important religious center for the three main monotheistic religions of the world.

The findings show that three factors, "length of stay," "main purpose of visit," and "number of visit," create a framework or spectrum of opportunities for the tourist. The way these factors tend to shape the spectrum of possibilities for the tourist reminds one of Lund School theories of time-space geography. In contrast to time-space geography, factors such as "gender," "age," and "income" do not bear a strong impact on the constraints themselves. It is clear, however, that the religion variable is of significant importance in Jerusalem.

Other variables that have yet to be identified and tested may also exist. Indeed, future research might investigate the effects of cultural curiosity, the tourist's image of the city, use of guidebooks, differences in personality structure, and other factors.

The findings of the study suggest that the "constraining" variables and the principles underlying the spatial activity of local residents and tourists in cities are very different. This stems from the fact that the activity of the local resident tends to be the outcome of long-term decisions, whereas the activity of the tourist tends to originate from short-term decisions. The day-to-day activity of the local resident is determined by decisions regarding career choices, selection of workplace, place of residence, and various other factors. Such decisions are largely influenced by the individual's education, age, gender, and income level. For example, factors such as gender and income, which are important in explaining the spatial activity of individuals in everyday life (Hanson and Hanson 1981), will be less meaningful in explaining the activity of these same individuals in a city in which they are visitors. In contrast to the local resident, the tourist is unencumbered by such long-term decisions because, as implied by his or her tourist identity, he or she will stay in the city only for a limited duration. The tourist does not need to take care of a household—this is taken care of by the staff of the hotel or, alternatively, by the family he or she is staying with. If children are taken along for the trip, they do not need to attend school or other educational frameworks and, therefore, the family is free to plan visits to tourist sites that will satisfy all members of the family or tour group.

Similar findings emerged in a study conducted by Debbage (1991, 266), which found that the major determinants of spatial behavior were temporal constraints (length of stay, mobility levels) and the spatial structure of the resort environment. However, Debbage found that some of the more traditional explanations for spatial behavior did not seem to be applicable in his case study of Paradise Island (Bahamas). For example, factors that were not significant included the socioeconomic characteristics of the individual (income, education, age, and occupation), the type of travel arrangement (packaged tour vs. independent traveler), familiarity levels (first-time or repeat visitors), and party size. Part of the reasoning for this may be because research in other fields (intra-urban commuting patterns, consumer shopping behavior, and residential location decisions) may not be directly transferable to tourist behavior (Debbage 1991, 266).

DIFFERENTIATING BETWEEN INDEPENDENT TRAVELERS AND ORGANIZED TOURISM

The clearest and most absolute distinction between types of tourists in terms of space-time activity (as well as in terms of sociology; see Cohen 1972) is the organizational structure of tourists during a visit. Individual tourists (known as fully independent travelers, or FIT) traveling independently are distinguished from organized groups. The differences between these two types stem primarily from the "rigidity" of the organized groups' itineraries in contrast with the greater freedom the individual tourist experiences when choosing sites to visit. As a result of these differences, the spatial models of these two groups differ to a great extent.

The differences in spatial activity are expressed in a number of ways. Organized groups move from place to place on buses while individual tourists' destination consumption is primarily done on foot. The individual tourist's visits at different locations are more comprehensive and in-depth. Chadefaud (1981), studying pilgrimages to Lourdes, found that individual tourists have a wider activity range than do organized groups, which are focused more in space. In Shoval's (2002) study conducted in Jerusalem, it was found that while intensive organized tourist activity takes place in all tourist locations in the city, within the locations themselves group activity focuses on one site while individual tourists visit other sites in the same location. On Har Hazikaron (the Mount of Remembrance), for instance, organized tourist groups focus on Yad Vashem, the Holocaust memorial and Museum, while individual tourists visit other sites in the area such as the military cemetery, Har Herzl (Mount Herzl), and Prime Minister Rabin's grave.

Organized Tourism

In the case of organized tourism, unlike with individual tourists, an itinerary is designed by a tour operator and therefore different attributes of individual tourists in the group are not relevant to the spatial patterns of the group. Due to the fact that the tour is essentially a contract between the tourist (the client) who purchases a tour package and a tour agent (the seller), it must be executed exactly as written, and the flexibility available to the tour guide or local agent is limited to changing the order of sites to be visited on a given day or on different days.

In groups that are formed on a religious, organizational, or social background, there is usually a "group leader," often someone who is a religious leader for the community (priest, rabbi, etc.) or responsible for the visit on behalf of an initiating organization. The group leader may at times express desires or preferences of his or her own in terms of the combination of different locations and activities included in the tour. In cases in which the group leader comes to an agreement with the agent about the financial significance of changes to the itinerary before the final price is settled, these may be included in the itinerary.

The travel agent has two principal concerns when building an itinerary. The first is his or her past experience; that is to say, itineraries that have proven successful in the past. An agent will always attempt to repeat tried and true formulas rather than invest resources in creating a distinct itinerary for each group.

The second criterion is budgetary: The agent must market his or her product to potential tourists at an attractive rate, thus it is worth his or her while to include locations that are open free of charge whenever possible in order to reduce the overall price of the package.

The critical significance of the travel agent in shaping an itinerary stands in contrast to Cohen's (1985, 14) view on the matter. Cohen suggests that, due to the fact that the guide has an important role in choosing and shaping the itinerary, his or her personal preferences and the training process he or she undergoes as a guide have an important place in shaping the itinerary, in addition to the desires of the group and the directions of the travel agent. As we saw, the tour guide is a "sub-contractor" of the tour operator and has no ability to change the itinerary he or she is given; once received, it is complete and ready for execution. Removing a site from the itinerary can only be done with the approval of the group leader and the agreement of the group. Adding a site is only possible if it will not incur additional expenses for the agent. However, the tour guide may be more flexible with the trip schedule, changing the order of visits, or interchanging days.

The Individual Tourist

The primary purposes of an individual tourist's visit to a destination are more varied than those of a tourist in an organized group. In addition, differences between the two types of tourists are expressed in the duration of their stay, their familiarity with the space, their perception of the space, and more.

The individual tourist has two principal forces dictating the nature of his or her visit. The first and most important is the time at his or her disposal while visiting a location. The longer he or she stays at the destination, the more sites he or she will be able to visit. The second factor is the purpose of the visit. The purpose may be seen, in some cases, as a factor influencing the tourist's spatial activity at the destination. For example, a business tourist who comes to a destination in order to promote sales is not free to tour the area as is an individual who came to see the sights. Other factors influencing tourist activity are the hours of activity of certain sites and their days of activity, the information the tourist has about the destination, and more.

These factors remind one of Hägerstrand's (1970) time geography (see the next chapter for the use of this approach for time-space visualization), mentioned earlier in this chapter.

Prior Perception and Familiarity

This is relevant only regarding individual tourism since, as was explained previously, the itineraries of organized groups are predetermined. The flexibility of the individual tourist means that the prior perception and/or the familiarity of the tourists with the destination are significant. Tourists always possess a perception of the destination prior to their visit. This perception is shaped by a combination of factors. On one hand, their perception stems from years of education and exposure to information about the location in literature, film, culture, and more—which make up a long-term influence. This is true primarily when the destination is a big city or a place of special importance. On the other hand, it is the short-term factors that affect the tourist's spatial perception and activity at the destination; for example, the tourist's familiarity with the space influences the nature of his or her interest in tourist sites. If the trip constitutes a first visit, the tendency is to visit a relatively greater number of sites, generally sites that are identified with the destination. As the tourist's familiarity with the space grows, there is a tendency to visit fewer sites, to visit relatively newer sites (which the tourist did not know about on earlier visits), and to visit less-well-known tourist sites. In other words, the tourist undergoes a transition from shallower to more in-depth tourism.

The tourist's perception about a given destination also stems from the targeted actions of tourist destinations, actions taken in order to draw attention in an effort to encourage investments and tourism. Cities, for example, attempt to create a certain image using various symbols; for example, symbols of industry ("motor city"), symbols of vision ("the city of tomorrow"), and symbols of high quality of life ("the city of leisure"). Slogans such as these are generally tied to images or objects, such as a picture of the horizon or some other type of graphic logo. In this way, a unique symbolic identity is created for a place, giving the impression that that place is special (Gottdiener 1994, 15).

It should be noted that the tourist in an organized group also comes on a trip already holding a perception of the space and a familiarity with the space on some level. However, as the itinerary is predetermined, the group tourist's perception is irrelevant to his or her spatial behavior, which is a result of the agent's interpretation of the tourists' preferences.

As many cities in the world are not considered to be foremost tourist cities, the tourist will have no clear perception of these before his or her arrival for a first visit. In these cases, the importance of different sources of information at the tourist's disposal before arrival increases. One might ask how these sources of information affect the shaping of spatial patterns for different tourists. For a detailed discussion of the variety of sources of information available to tourists before and during their trips, see Dann (1996, 135–170).

An important source of information at the tourist's disposal is the guidebook. It serves the same function for the individual tourist as the tour guide does for the organized group. However, the influence of the guidebook will be less pronounced than the influence of a tour guide on an organized group. The dissertation of Lew (1986; 1987) on tourism in Singapore confirms this assumption, demonstrating that there is no clear connection between guidebooks and the spatial activity of individual tourists in the city. Similarly, the guidebook allows the tourist to supplement essential details about his or her visit to the city, as in actuality, due to the size of present-day cities and due to the short duration available to the tourist, he or she will not be able to tour the entire city.

Accommodation

An additional difference between individual tourists and organized groups is seen in their choice of accommodations. The model suggested by Yokeno (1968) for explaining the locations of hotels in space uses the idea of economic rent to explain financially why individual tourists prefer to choose hotels located specifically in the center of town. This model was empirically proven in a number of other studies conducted later on (Egan and Nield 2000; Shoval 2006).

This model is relevant only for individual tourists since members of organized tour groups usually have a tour bus, tour guide, and bus driver at their disposal. The tour bus enables tour groups to move effortlessly throughout the city, while the presence of a tour guide and bus driver, both of whom are familiar with the city's highways and byways, means that, unlike the individual tourist, the group has no difficulty orienting itself in the city.

It is evident that regardless of the type of city under discussion, the majority of tourist accommodation is to be found in or around the city center for the simple reason that tourists, generally, want to be either in or within walking distance of the "Tourist City." While Yokeno's model offered an analytical explanation as to why hotels cluster in the inner city, as regards to the question why there is such a great demand for centrally located hotels on the part of tourists, Yokeno was content to note that: "The large concentration of hotels in the city centre enables the individual tourist to choose a hotel 'according to his taste and budget'" (Yokeno 1968, 16).

Thus, the fact that they are clustered in a small geographic area gives hotels an important marketing advantage. In our view, however, there are four additional suppositions that may be added to Yokeno's assumption, further shedding light on tourists' preference for hotels in the city center and reinforcing the Yokeno model's contention in regards to the desire of individual tourists for a hotel close to the city center and their willingness to pay for the privilege:

1. Tourists prefer to stay in centrally located hotels so as to reduce the time spent traveling to and from the city's tourist center, thus maximizing their stay in the city (Cooper 1981).
2. Not usually in possession of a car, tourists want to be able to reach the city's tourist sites by foot. There is, as a rule, little point in them renting a car since they are rarely familiar with the city's road system, particularly when it comes to big cities. Parking in city centers is often expensive (Burtenshaw et al. 1981, 172). This assumption depends on the individual city's spatial structure and on the reason for the visitor's sojourn in the city. In cases in which visitors to the city have a car at their disposal and no reason to go to the city center, then obviously a hotel located on the edge of the city is preferable.
3. Given their limited knowledge of the city's layout, tourists prefer to reside close to the city center where most of the tourist attractions are located. Not only do they find it difficult to orient themselves in the city, but, because their spatial mental map is usually fragmented, tourists find it difficult to calculate geographical distances, including, for example, the distance between different hotels and the city center (Burtenshaw at al. 1991, 211).
4. Even when able to calculate geographical distances, most tourists are not familiar with the city's transport system, though they are vaguely aware of what the best route and method of getting to different locations is; nor are they cognizant of the time it takes to travel from one place to another (Arbel and Pizam 1977).

When summarizing the importance to tourists of quick and easy access to the city center, Pearce noted that:

A central city hotel responds to the needs of a variety of urban visitors whose stay is generally relatively short, typically in the order of two to four days. Many business travelers value proximity to the Central Business District (CBD) so as to transact their affairs promptly and efficiently. Here too the shopping-oriented visitor can descend rapidly on a range of high order shops. It is also in the centre that the sightseeing tourist will find many of the city's historic buildings, monuments and other cultural attractions. The overlapping of the focus of activity of each of these different markets reinforces the concentration of hotels in the central city. (Pearce 1995, 151)

The Case of the Hybrid Individual-Group Tour

One "hybrid," combining organized tourism with individual tourism, exists in the form of the cruise tourism. This is organized tourism in the extreme, on one hand, as the majority of time is spent in a relatively narrow area, with activities highly regulated by a tour operator. However, when the cruise ship reaches certain ports and "dismisses" the tourists for a brief period of time, if they have not purchased the organized package for the time to be spent in the location, they will function as day-tourists.

Another example of an individual-group hybrid is mass tourism at vacation villages close to the beach. While this type of tourism functions in a very foreseeable and constant manner, at times tourists will take advantage of their stay to visit tourist locations nearby and then their activity at the destination will be more akin to that of individual tourists. One of many examples of this phenomenon is the tourism at vacation towns in Costa Dorada (Catalonia, Spain), like Salou or La Pineda, who travel to Barcelona for a day independently while staying there.

CONCLUSION

This chapter identified a number of factors that determine the spectrum of opportunities available to tourists. Those opportunities shape their spatial activity in the destination. Additional characteristics that may be of importance in explaining the spatial behavior of tourists, as distinct from the ordinary activity of the residents in the city, should also be investigated, since, as was presented earlier, the spatial activity of residents of a city stems from entirely different factors and considerations than does the spatial activity of the tourist.

The identification of factors that influence the spatial behavior of tourists enables the elaboration of a typology that is based on the spatial activities of the tourists. (This will be discussed and demonstrated in detail in Chapter 7.) Such a typology will be fundamentally different from other typologies in tourism research (for example, Cohen 1972; Plog 1973; Cohen 1979), which tend to be a-geographical by nature and, therefore, not useful for the spatial analysis of tourism in cities.

3 Methodological Aspects of Measurement and Visualization of Tourists' Spatial Behavior

The time-space activity of tourists in a destination is of great interest for tourism researchers, but it can be very difficult to record. This chapter presents the primary methods that were used for data collection of tourists' time-space activities prior to the introduction of tracking technologies as tools for the same purpose.

As one would expect, a large number of the data collection and visualization methods are drawn from geography. Geographers were preoccupied with questions of movement in space for decades before tourism studies developed; when the area of tourism geography was created and began to develop, the interest in tourists' time-space activities began to emerge as well, borrowing tools and methods from geography. Indeed, many tourism researchers studying issues related to tourist activity in time and space were trained as geographers, the authors of this book among them.

DATA COLLECTION

As was evident in the previous chapter, current tourism theories are not very detailed in terms of their temporal and spatial resolution. This is probably due to the low accuracy of the existing data collection methods implemented in tourism research.

Two types of methods are currently employed by tourism researchers to gather information on the temporal and spatial behavior of tourists: direct observation techniques and non-observational techniques.

Direct Observation Techniques

Methods of direct observation can be summed up using the words "identify, follow, observe and map" (Thornton et al. 1997, 1851).

Participant-Observer Method

The first method of observation, known as "participant-observer method," involves the observer accompanying the individual under scrutiny in person. This approach is widely used in anthropological research when the researcher aims at achieving an intimate familiarity with the research population.

Non-Participatory Observation

Alternatively, the observer may follow the subject(s) at a distance, recording the pattern of their activities over time and space. This technique is known as "non-participatory observation."

When studying the spatial and temporal behavior of American tourists in Munich, Hartmann (1988) used both techniques, but was happy with neither. Both, he noted, were incredibly time-consuming. Nor were these the only problems Hartmann encountered. While the non-participatory technique yielded a mine of information, it failed to unveil the purpose and meaning underlying the subjects' decisions and activities. It also posed various ethical questions, particularly when pursued in covert form (Hartmann 1988, 94–101). While these were less of a problem when it came to the participant-observer procedure—the observer, thanks to his or her intimate contact with the subjects, was constantly aware not only of what the subjects were doing but also, possibly, why—there was in this case the risk of the subjects tailoring their behavior and explanations, albeit subconsciously, to the presumed expectations of their observer-companion. In addition, this method proved very expensive, as the researcher had to be present for the entire duration of the tourist's exploration of a destination.

Keul and Küheberger (1997) used the non-participatory technique in order to analyze the spatial behavior of tourists in Salzburg, Austria. Hoping to resolve the problem of uncovering motivation as well as to accurately trace parts of the participants' routes, the two followed up their non-participatory observations with a series of interviews of the tourists observed. However, as the non-participatory technique can be applied only for very short time periods due to the high costs of the method, observations were limited to fifteen minutes out of what were, on average, four-hour-long walks. Murphy (1992), who conducted a similar non-participatory-type observation study in the city of Victoria, British Columbia, kept his subjects under surveillance for an average of twenty-three minutes, the longest surveillance period lasting eighty-seven minutes and the shortest four minutes (Murphy 1992, 206).

This problem intensifies when the destination visited is not a historic city, but instead a location characterized by relatively long stays, such as a large multifunctional city, seaside resort, or skiing area, where tourists might spend as long as an entire week. Another drawback to this technique is that beyond a simple visual estimate as to the subjects' socioeconomic background, it cannot, in those cases in which interviews are not used, render non-visual data such as the total length of visit and the subject's next destination. The method is thus not only hugely expensive, time-consuming, and labor-intensive; it is also not comprehensive and does not gather all data necessary on the subjects.

Recently, researchers from the University of Girona (Gali-Espelt and Donaire-Benito 2006) published a large-scale study of the time-space activities of visitors to the Old City of Girona (Catalonia, Spain). In order to learn what the main sites of interest were for the visitors ("nodes" of activity), what the length of stay of visitors to the Old City was, and what the total distance walked by the visitors (in meters) was, a total of 532 individual tourists were tracked for just over a year (July 2002–September 2003). This time frame was determined in order to take into account the seasonality of tourism in Girona, which is characterized by two intense periods (spring and summer) and two quieter periods (autumn and winter). Since there are various entry points to the Old City, several sampling points were chosen. The visitors were observed by a team of two persons who followed them (one hopes) unnoticed, recording their temporal and spatial activity and adding comments about the behavior of the participants using voice recordings. At the end of their visit, shortly before the visitors left the Old City, the followers identified themselves and gave the visitors a questionnaire to complete.

Remote Observation

A less costly non-participatory technique is remote observation, which is used to record and analyze aggregate tourist flows. Hartmann (1988) used this technique in Munich, where he positioned a camera on top of the city hall's eighty-meter-high spire and took aerial pictures of the crowds gathering below to watch the Glockenspiel in the Old City's main square. He then used the pictures to estimate the percentage of young North American tourists among the total number of people watching the ten-minute display. The question remains whether the identification of the North American tourists by their appearance alone was entirely accurate.

The method described above had already been used several decades earlier to track the routes selected by pedestrians through a busy office parking lot, in this instance by setting up an observation point in a tall building overlooking the parking lot (Garbrecht 1971).

Aggregative Video Tracking

Recent years have seen a development in the ability to use data obtained by video cameras or closed-circuit television (CCTV) cameras arranged in public places to analyze behavioral patterns of users. Many municipalities have installed networks of CCTVs that cover a large area within their territory; they do this in order to monitor the activity under their jurisdiction in real time. The idea is not new, but until recently the measurement was done using costly manual processes, such as behavioral mapping and time-lapse filming with human examiners. In recent years, applications of computer vision enabled these processes to become automated and therefore relevant for aggregative analysis of mass data on human behavior in public spaces (Yan and Forsyth 2005). For example, the analysis of the number of people present in a specific location in the city at different times during the day could be very useful in assessing the physical carrying capacity of a specific place; or, if various areas in the city are monitored by CCTV cameras, visitor volumes at different points of interest in the city may be calculated.

Obtaining surveillance tapes is a barrier that stands between the researcher and this source of data. The information on the tapes belongs to the authority that positions the cameras, usually the municipal authority or police. These agencies might not be willing to make the data collected accessible for research purposes.

Hill (1984) has noted, that all of these "eye in the sky" techniques, though effective for studying the behavior of individuals within a restricted spatial setting, are of little use once the pedestrians step beyond the observation point's line of sight (1984, 542). This criticism is true of most other fixed-point observation studies, whether using time-lapse photography, video recorders, or the increasingly ubiquitous closed-circuit television (Hill 1984; Helbing et al. 2001) as described above.

Cellular Data

In Chapter 8 of this book, a methodology of using aggregative data obtained from cellular network providers will be presented. Although this concept can yield interesting theoretical findings and practical knowledge, it has the same weaknesses as the remote observation techniques described above. While cellular phones can provide an objective snapshot of the subjects' behavior, they cannot, by their very nature, reveal the motivations underlying the activities thus documented (Hartmann 1988, 100). This is the technique's principal drawback, one that it shares with other non-participatory procedures.

Non-Observational Techniques

Time-Space Budgets

The time-space budget technique is the most common method used for collecting data on human time-space patterns in the social sciences in general and in tourism studies in particular. Time budget and time-space budget are defined as a systematic record of a person's use of time over a given period. It describes the sequence, timing, and duration of the person's activities, typically for a short period ranging from a single day to a week. As a logical extension of this type of record, a space-time budget includes the spatial coordinates of activity location (Anderson 1971, 353).

This method is non-observational and relies on the subject's report of his or her behavior. Utilizing one of four procedures—recall of a specific time period, recall of a "normal" time period, game playing, and time-space diaries—it records human activity in time and space, over a limited time period, usually between a day and a week (Parks and Thrift 1980). Of the four procedures used, the time-space diaries have a distinct advantage over the rest in that they can record behavioral patterns that are impossible to observe directly owing to their spatial and/or temporal nature (Thornton et al. 1997).

However, all four procedures have a methodological flaw, since they all require the subject to actively record his or her actions in time and space during the entire experiment. In other words, they are based on the subject's collaboration, something that undoubtedly affects the quality of the data gathered. While there are examples of substantial success using this methodology (see, for example, Goodchild and Janelle 1984; Janelle et al. 1988), the burden placed on prospective subjects also explains why there is some difficulty in finding people willing to participate in research using this methodology and why those who do often fail to faithfully record their activities (Szalai 1972).

Time budget methods were first developed in the nineteenth century within the framework of surveys studying the standard of living of working-class individuals. The slogan "eight hours labor, eight hours recreation, eight hours rest" adopted by workers' unions at the turn of the century expressed a social demand in the form of a laconic time budget (Szalai 1972).

Geographic research began to use these methods primarily from the 1960s (Anderson 1971, 353). It should be noted that the studies conducted at the time did not relate to tourists as active figures within a city. The tourist field did not fully realize the great potential existing in time diaries at the time, and therefore it is possible to name only a few studies that used this method (for example, Murphy and Rosenblood 1974; Cooper 1981; Lew 1986; Pearce 1988; Debbage 1991; Dietvorst 1994; Thornton et al. 1997; Knaap 1999).

Fennel (1996) notes that among the tourism researchers using time budget techniques only a few were geographers, and therefore the studies focused more on the questions of "why" and "what" than on the question of "where." Among the studies of tourists' spatial behavior, the majority focused on vacation areas rather than multi-functional urban areas. This can be explained by the fact that there is a methodological advantage to conducting research in locations in which tourism is the primary activity and in which tourists have a limited number of purposes for their visits. Pearce's (1987) cynical comment that the majority of studies in tourism focus on vacation locations due to the preference of tourism researchers to conduct fieldwork in those areas also comes to mind.

Based mostly on time-space diaries, the time-space budget technique records behavioral patterns, which, owing to their spatial and/or temporal nature, are impossible to observe directly (Thornton et al. 1997). Other than this all-important asset, the technique has all of the advantages of questionnaire-type surveys.

Of the various studies employing this technique, the most notable are Murphy and Rosenblood's (1974) work on the spatial activities of tourists and daytrippers in British Columbia's Vancouver Island; Cooper's (1981) investigations into the behavior of tourists in the Channel Islands; Lew's (1987) study of tourist spaces in Singapore; Pearce's (1988) analysis of the spatial behavior of tourists in the island of Vanuato in the South Pacific Ocean; Debbage's (1991) examination of tourists' behavior in Paradise Island in the Bahamas; Dietvorst's (1994; 1995) research into the time-space activities of tourists in the small historic city of Enkhuizen in Holland; Fennell's (1996) work, based on time-space diaries, in the Shetland Islands; Thornton et al. (1997) investigations into the spatial behavior of tourists in the Cornish resort of Newquay; and Zillinger's (2007) research on the time-space activities of German car tourists in Sweden. This impressive body of work apart, most researchers in the field have tended to avoid the time-space budget method, owing to its various shortcomings, most of which center on the difficulty of keeping accurate records (Hall and Page 2002, 46).

In practice, the time-space budget procedure utilizes one of several techniques. The first involves recall diaries, usually in the form of a questionnaire or interview, which the subjects complete post factum. Both Cooper (1981) and Debbage (1991) employed recall diaries during their work.

The principal problem with recall diaries is that the amount and quality of information gathered depends on the subjects' ability to recollect past events with any degree of precision and detail. Furthermore most questionnaires are, of necessity, phrased rather succinctly lest the subject lose patience, which inevitably limits the amount of information obtained. Face-to-face interviews, on the other hand, while allowing for more detailed questioning, are again dependent on the subjects' memory; however good the subject's memory might be, most people will have only the haziest notion, or at best an approximate idea, of the frequency, sequencing, and duration of their activities.

Self-administered diaries, to be filled by the subject in real time, make up another time-space budget technique. Researchers use such diaries when studying the spatial and temporal behavior of tourists (Fennell 1996; Lew 1987; Pearce 1988; Thornton et al. 1997). Recently, Connell and Page (2008) used a self-completion map-based questionnaire to identify aggregate travel flows of car-based tourist travel in the areas of Loch Lomond and Trossachs National Park in Scotland.

However, while resolving better the question of memory lapses, these diaries have several problems of their own. Above all, they demand a considerable effort on the part of the subjects, who are required to record in detail their spatial activities while busy enjoying themselves touring the city or countryside. It is a distracting, disruptive, tiring, and time-consuming process, which goes far in explaining why so few are willing to take part in such studies. In addition, this method has low resolution in terms of time and space, since the research subject records his or her location and activity just several times per day. Even if a subject agrees to record his or her whereabouts every thirty minutes, the resolution of the data obtained is very low in comparison to the new tracking methods that will be elaborated later in this book.

Table 3.1 provides an example of a page of an activity diary designed for recreating activities at the end of a visit to a destination. The time resolution in this case is every three hours throughout the day. This is a very low resolution, but customary in many tourism studies. It does not allow the examination of exact durations of activities but identification of main patterns in time and space. This approach, dividing the day into five or six time slots, is characteristic of earlier studies of tourist activity in time and space, research which was based on gathering data using time diaries (Cooper 1981; Debbage 1991; Thornton et al. 1997).

Yet, even among those ready to volunteer their services, there will be distinct differences in terms of their commitment and enthusiasm, and consequently considerable variations in the quality of the information obtained. Moreover, the longer a project goes on, the less keen and cooperative most subjects will become (Pearce 1988, 113), leading to a sharp fall in the amount and quality of the data recorded (Anderson 1971). According to Pearce (1988, 116) a week is the most that one can expect a subject to compile such a diary in any satisfactory or meaningful manner.

Table 3.1 Sample Page of an Activity Diary

	From the exit of your hotel until noon	12–3 P.M.	3–6 P.M.	6–9 P.M.	9 P.M.–until back in your hotel
Day 1	....	....	....	....	....
	....	....	....	....	....
	....	....	....	....	....
	....	....	....	....	....
	....	....	....	....	....
	....	....	....	....	....
	....	....	....	....	....
	....	....	....	....	....
	....	....	....	....	....

Self-administered diaries completed by the subject in real time are no doubt more accurate than the ones filled retrospectively at the end of a visit. However, the following example presents a case from a study outside the realm of tourism research, but still of high relevance, clearly demonstrating the limitations and potential methodological flaws with the self-administered diaries method. The example is taken from "SenTra," an interdisciplinary project that focused on the time-space activity patterns of urban-dwelling, cognitively impaired persons, as well as unimpaired controls, over a period of one month (Shoval et al. 2008). A tracking kit including a GPS receiver and a device resembling a wristwatch was distributed. The combination of the two devices allowed the researchers to know whether participants were carrying the GPS device at a given moment. The upper part of Figure 3.1 presents some of the daily activity of one elderly participant according to his self-administered diary, which was completed with the help of his spouse. According to the participant, he visited a cemetery at a distance of 1.5 kilometers from his home for about one hour in the early evening on September 7, 2008. This activity was validated by the tracking data obtained by the tracking kit (see the lower part of Figure 3.1). However, two days later, on September 9, 2008, the accurate data obtained from the tracking kit gave evidence of yet another visit to the same cemetery at a similar time. This event was not recorded in the activity diary that the participant and his wife completed (again, see the table in Figure 3.1). When interviewed at the end of the participation period, the couple claimed that they had never left their home that evening. Due to the nature of the tracking kit, there is no doubt that this couple did visit the cemetery that evening. This example illustrates the problematic nature of the method of self-administered time-space diaries. At the same time, however, it also expresses the advantages of using new available tracking methodologies that can support or replace the traditional methods.

The different advanced tracking technologies that will be introduced in the next chapters of this book have the abilities to resolve some of the problems that were discussed regarding the current methods used in tourism research to collect time-space data about tourists. Collection of objective time-space data with high resolution in time (seconds) and space (meters) for long periods of time opens up the possibility of drawing new lines of inquiry and creates opportunities to formulate new research questions that could not be asked previously. Naturally there are also limitations and challenges with those new technologies. These will be discussed in later parts of this book. Finally, it should be emphasized that the adoption of new tracking technologies does not mean that traditional tools such as interviews, questionnaires, and time-space diaries must be excluded from tourism research in our view. On the contrary, the new technologies will add to the existing ones and will allow the synergic use of new and old methods. Examples of the combination between the traditional and the new technologies will be presented in Chapters 7 and 8.

Date	Starting time	Finishing time	City	Street	Location according to GPS	Location according to diary
Sep.7.2008	17:23	18:35	Herzliya	Pinsker	Cemetery(15)	Cemetery
	18:38	18:46	Herzliya	Pinsker	Entrance to Cemetery(16)	Cemetery
Sep.9.2008	17:30	18:25	Herzliya	Pinsker	Entrance to Cemetery(16)	Home (1)
	18:25	18:47	Herzliya	Pinsker	Cemetery(15)	Home (1)

Figure 3.1A Record from activity diary kept by a research participant.

DATA VISUALIZATION

Two strategies exist for dealing with the analysis of time-space data. One is to "chop the data," or, in more academic words, to categorize the spatial and temporal data so that variables can be created. The advantage of this strategy is that it enables the merging of the geographical data with the socio-demographic and other attributes of the tourist and the environment; this allows for all kinds of well-known statistical procedures. The other strategy is to treat the time-space activity as "one unit," borrowing mainly from Hägerstrand's time geography concepts. This enables the researchers to learn about the temporal sequence of events, but to keep the spatial

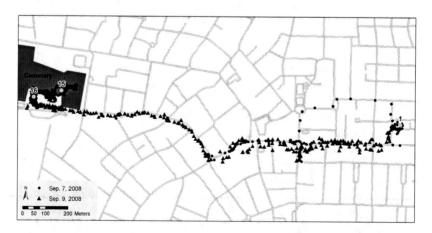

Figure 3.1B Plotted GPS track of research participant.

aspects of the activity. This strategy has gained impetus in recent years in geography in its broader sense, and has also flourished in tourism analysis (see Forer 2002, for example) due to various conceptual contributions and practical developments in the field, as will be presented later in this chapter and in Chapter 7 as well.

In this chapter, we chose to focus on the latter strategy for two main reasons: Firstly, textbooks already deal with the issue of statistical analysis and tourism research and secondly, we believe that analyzing time-space data while maintaining the temporal sequence of activities and geographical context is the better option to choose when high-resolution data collection and analysis technologies are available. This enables researchers today to perform analyses that were not possible only a few years ago.

The Visualization of Space-time Paths

Hägerstrand's concept of time geography (1970) captured the way in which the spatial and temporal characteristics of people's activities constrain them in their day-to-day lives. As time geography has it, human activities occur in specific locations and for limited time periods only. Thus transport systems, by enabling people to travel from one place (or activity) to another, allow them to make more efficient use of their space-time restrictions by trading time for space.

One of Hägerstrand's most profound professional and disciplinary achievements was the ability to represent space and time in a single diagram (Gren 2001) unlike an ordinary map, but rather like a snapshot, reproducing a moment frozen in time. The result was his now famous time-geographic diagram, a notational (representational) system, which forms the basis of much of the subsequent work in the field of time geography, particularly in the realm of analysis and interpretation. The diagrams (Figure 3.2, taken from Gregory 1989) consist, as a rule, of three axes, a time axis and two space axes, making it possible to trace individual time budgets in graphic terms. The effective range of each person is described by a prism, or a series of prisms, whose shape is dependent upon the constraints, as we saw earlier. Hence, every pause, regardless of the activity involved, will cause the prism's (or sub-prism's) range to shrink in direct proportion to the time spent at said stop. But there are also other wider structural features, specific to the social systems within which individuals operate, which, as has long been recognized, help shape people's time budgets and activity patterns.

Pattern Aggregation Methods

Hägerstrand's time-space aquariums presented previously serve as a strong visual tool but do not have a quantitative basis for aggregation behind them. Rather than apply quantitative methods to the observed spatial data,

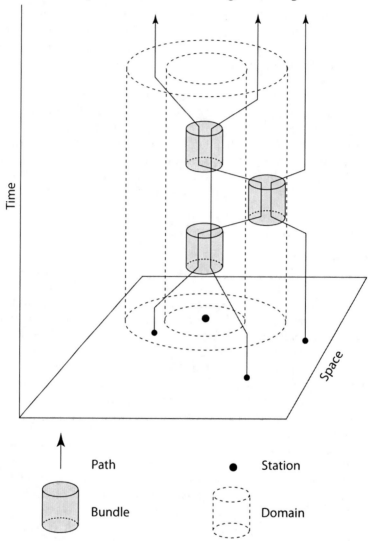

Figure 3.2 Hägerstrand's time-geographical diagram. Source: Gregory 1989.

early time-geography studies used time-space diagrams to represent what the researcher thought to be a typical pattern of behavior of a certain group within the population (see, for example, Palm and Pred 1974); these patterns were developed by intuition rather than by using a set method that would help the researcher understand and categorize the different observations. Several efforts have been made to develop methods to divide and categorize a spatial-temporal data set into groups based upon spatial characteristics within the data set. These efforts can be divided into two main groups: quantitative pattern aggregation methods and visual pattern aggregation methods.

Quantitative Pattern Aggregation Methods

The main problem researchers encounter when trying to assess similarities between spatial patterns is the lack of an accepted methodology that measures similarity (Schlich and Axhausen 2003). One of the main factors preventing agreement upon one accepted method that measures similarity in time-space data is the difference in views regarding the importance of the various components that make up time-space data (such as duration, location, and nature of activity). These varied views lead to different methods of calculating similarity and hence to different results (Huff and Hanson 1990).

Some examples of attempts to aggregate behavior patterns using quantitative methods include counting the number of trips made (Pas and Koppelmann 1986) and grouping different characteristics of the trip such as duration and purpose (Huff and Hanson 1986). None of these attempts effectively developed a method to measure similarity while taking into account the sequential aspect of the time-space data, as was pointed out by Schlich and Axhausen (2003) in the conclusion to their paper:

> Still, the methods have disadvantages. For example, they do not take the order of the activities into account. Further methodological research is needed to incorporate the full information that is available in a travel diary. One promising approach is the sequence alignment method (Wilson 1998; Joh et al. 2001) that measures similarity based on the Levensthein distance (Sankoff & Kruskal 1983) instead of an Euclidean measurement. Unfortunately, the sequence alignment method is more complex than the methods analysed in this paper and requires a high level of abstraction for its interpretation. (Schlich and Axhausen 2003, 34)

Implementation of a sequence alignment method for analyzing time-space patterns will be demonstrated in the second part of Chapter 7.

Visual Pattern Aggregation Methods

Forer (1998) and Huisman and Forer (1998) were among the first to create time-space prisms using a geographic information system, or GIS. In their work they implemented space-time paths and prisms based on a three-dimensional raster data structure (for an explanation of this term, please see the appendix) for visualizing and computing space-time accessibility surfaces. The use of a raster structure is not suitable for the depiction of transportation systems, the infrastructure upon which most spatial behavior takes place. Kwan (1999A), in her study of non-employment activities, presented the time-space data obtained in three-dimensional time-space prisms or aquariums using vector GIS methods (Figure 3.3). These methods have turned out to be more efficient and more suitable for use in the creation of three-dimensional time-space prisms with GIS software (Kwan 2000).

As discussed above, since their introduction, Hägerstrand's time-space prisms have served as a prominent tool for the study of behavior in time and space, but for many years after their introduction, the production of time-space prisms was limited to manual methods. These methods were labor-intensive, involved the assistance of a draftsman, and resulted in one static view of the prism. Adding another path or changing the view meant returning to the drafting table and restarting the process from the beginning. Today's sophisticated geographic information systems allow for the relatively easy and quick plotting of time-space prisms and for interactive visualization of the data presented. Another feature of GIS systems is the ability to combine several views or data windows into one image (for example, a two-dimensional and three-dimensional view), enabling a fuller

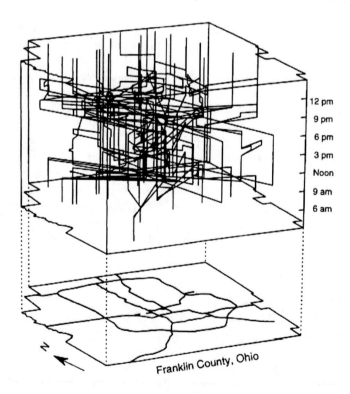

Figure 3.3 Space-time aquarium for women employed full time. Source: Economic Geography, 75, 4, 1999, 370–94. Copyright (Kwan 1999). Reprinted with permission of John Wiley & Sons, Inc.

understanding of the time-space data. The most important ability that the new computation abilities provide is the opportunity to visualize the temporal dimension by using dynamic figures that move in time.

Quantitative methods, as discussed above, require the discretization of time-space data in order to enable the creation of measurable variables. This problem is especially severe in time-space data; both time and space are continuous and therefore need to be discretized, leading to the analysis of two discretisized variables. Geovisualization can assist in alleviating this problem since it allows for the exploration of the time-space data without the data being discretisized (Kwan 2000). Additional advantages of using computerized GIS systems to create time-space prisms are the ability to integrate large amounts of data from different sources, allowing the creation of complex and realistic representations of the urban environment (Kwan and Weber 2003), and the added three-dimensional functionality such as "walk-throughs" and "fly-bys" which allow the researcher to create a virtual world that has the feel of a real world (Batty et al. 1998).

Time-Space Models

Since being introduced in the early 1970s, time geography has led geographers and transportation researchers to develop models describing human activity in time and space (Timmermans et al. 2002). The objective of all of these models is to create an accurate representation of the spatial behavior of people in different time frames and contexts. Spatial behavior models are widely used in transportation studies and are especially useful when trying to estimate the impact that different transport regulations and systems will have on the transportation network as a whole. These models can be divided into two groups based on the number of factors that are taken into consideration when building the model: single-component models and multiple-component models.

Single Component Models

1. Activity duration and time allocation. These models view time as a resource that each person allocates differently. In early years these models were mostly descriptive, illuminating the different ways in which people choose to allocate time to the different activities in which they participate, such as the ratio between traveling time and the duration of different activities. In the past fifteen years statistics techniques (for example, logit models, structural equation models, and log normal models) have been increasingly used to model the allocation of time to different activities. (For examples of these models see Kitamura et al. 1988, Bhat 1996, and Yamamoto and Kitamura 1999.)

2. Departure time decisions. These models are concerned with the decision-making process that leads to the time a person chooses to depart

to a certain activity (usually work). These models have been especially important to transportation planners who are concerned with spreading departure times in order to relieve congestion. (For examples of these models see Mannering 1989 and Kroes 1990.)

3. Trip-chaining and stop-pattern formation. These models have been of great interest to both urban planners and transportation planners. Urban planners use the models to assist them in locating different land uses in a way that allows for optimal trip chaining and the combination of a maximum number of activities in one trip. Transportation planners use these models to seek ways in which to encourage activity chaining to a maximum. These models are especially interesting in the context of our research because of the importance of the sequence of activities in building the models. Sequence alignment methods that have been used to understand the different sequences of activity that people practice. (For examples of these models see Kumar and Levinson 1995 and Timmermans and van der Waerden 1993.)

Multiple Component Models (Agent-based Models)

These models try to emulate the complexity of human life by separating the complex problem into different modules. These modules are a collection of autonomous decision-making entities called agents. Each agent consists of an assessment of an individual situation and the decision made on the basis of a set of rules; the model as a whole simulates the interaction between the different agents. This way of thinking and simulating assists in breaking down the complexity of human nature and circumstances into smaller and simpler bases of behavior; the combination of these agents strives to emulate humans' complex existence.

1. Constraint-based models. These models study how the different constraints (such as opening hours of stores and institutions and the physical ability to move from place to place) lend to the schedule. In these models, schedule optimization algorithms are applied to generate all possible schedules when taking into consideration the different constraints of time and space. The development of these models has been led by geographers who have developed different combinatorial algorithms. One of the first models belonging to this group was the PESASP model developed by Lenntorp (1976).

2. Utility-maximizing models. These models serve as one of the foundations of discrete choice models (logit models). The basis of these models is the assumption that the individual will choose a combination of utilities (or, in our case, activities) in a way that will maximize the benefit he or she derives. The models measure the utility of each combination of activities by using a simple algebraic rule and choosing the alternative that scores the highest. When developing

these models, it is of pivotal importance to ensure that all of the alternatives calculated are viable. In order to guarantee this, nested logit models were developed. These are models that group similar activities that can be swapped for each other in groups and in a way that ensures that the choice is made between similar activities.

3. Computational process models. These models make decisions between activities that an individual chooses based on context-dependent heuristics. Because these models follow complex rules, they need strong computational abilities to simulate the choice process. The first model of this group was "Scheduler," developed by Gärling et al. (1989) and later applied to predict the impact that telecommuting would have on commuters (Golledge et al. 1994). Additional models include "Amos" (Pendeyala et al. 1998) and "Albatross" (Arentze and Timmermans 2000).

The limitation of all of the presented methods of space-time visualization and analysis is their inability to aggregate space-time paths to create generalized types composed of varied activities in order to create patterns fashioned on a quantitative basis while taking into account the sequential element. A solution to this shortcoming will be presented in detail in the second part of Chapter 7.

Part II

Available Tracking Technologies

4 Land-based Tracking Technologies

Land-based tracking systems are local tracking systems, featuring a series of antenna stations, also known as radio frequency (RF) detectors, distributed throughout a specific area. Land-based tracking systems are predicated on the principle that electromagnetic signals travel at a known speed along a known path (Zhao 1997). The land-based antenna stations, having received the signals sent out by the end unit, will, using various techniques, calculate the end unit's position.

Land-based tracking systems were the first modern technologies made available as navigation aids. At a time when sending satellites to space was futuristic and inconceivable. Electronic navigation systems were used by mariners and pilots in need of navigation assistance. The implementation on a grand scale of satellite-based systems for navigation (see next chapter) made some of the land-based systems for navigation obsolete about a decade ago. However, the massive increase in cellular phones using land-based radio systems and employing the cell sector identification (CSI) method to locate a cellular phone has made the tracking of tourists via cellular phone, both of individuals or in aggregate, a very attractive option (see Chapter 8 for more detail). This chapter describes the development of terrestrial radio technology and the technology behind it.

Numerous navigation systems, both land and marine, are based on terrestrial radio technologies, their accuracy ranging from about thirty meters to a few kilometers; the majority have, so far, been used as maritime navigational tools. The past decade has seen a marked rise in the technology's precision.

The technologies used in land-based tracking systems can be divided into a few main categories:

1. Technologies that use the time that it took a signal to travel as a measurement of distance;
2. Technologies that use the angle in which the signal is received in order to calculate the direction to which the unit can be traced; and
3. Technologies that rely on the system's network layout to locate the end unit.

TIME OF ARRIVAL (TOA) TECHNOLOGY

Time of arrival (TOA) technology is based on the fact that radio-electric signals travel at a known speed, the speed of light. This makes it possible to calculate the distance that the signal has traveled, given that the researcher knows the exact time the signal was emitted and the exact time it was received.

However, keeping track of time accurately requires the use of atomic clocks, a costly addition to base units. Thus it is not cost-effective to implement TOA technology in land-based location systems with many stationary antennas that give a service that is local and not global and rather is applied in the Global Positioning System where the number of satellites is smaller in proportion to the service area.

One application of the principles behind TOA technology is in the time difference of arrival tracking method.

Time Difference of Arrival (TDOA)

TDOA systems are based on a series of three or more land-based antenna stations that pick up transmissions from end units. By calculating the difference in time that it took the signal to travel from the end unit to the different stations, it is also possible to work out the difference in distance between the end unit and the stations, defining a hyperbolic curve on which the unit can be located. This information is then passed on to the system's central station, which establishes the end user's position by determining the convergence point of the readings of at least three stations. TDOA is also referred to as "hyperbolic positioning." Figure 4.1A illustrates the process of locating an end unit with a TDOA system.

Positions can also be obtained by end units if stationary antennas actively transmit a signal rather than passively waiting to receive signals from the end units. In this case, the end unit receives the signal from several stationary antennas and then its location is calculated.

Many different, mostly local, TDOA systems have been deployed over the years. In this chapter we mention only the most prominent among them. The first TDOA system was implemented in World War II when the British launched the DECCA navigation system. The DECCA navigation system first served the Allied forces, which needed a way to accurately land their ships on the coast. After the war, the system continued to serve mariners and was used widely by fishing vessels as well as helicopters landing on serviced oil platforms in the North Sea. In the year 2000 the system was shut down, leaving its users to turn to satellite-based systems like the Global Positioning System as navigation aids.

The OMEGA navigation system was another one that employed TDOA technology. This system, the first to enable navigation worldwide, was initiated by the United States Navy but was made possible by the cooperation

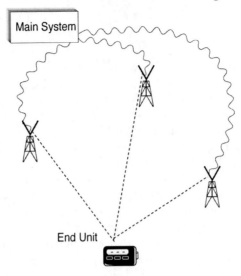

Figure 4.1A Location using TDOA technology.

of six different nations hosting transmitting stations all over the world. Using very low frequency (VLF), which allowed the signal to travel for long distances over the surface of the earth, and consisting of eight transmitting stations, this system, like the DECCA navigation system, was used by mariners (Appleyard et al. 1988). The system covered the whole globe, enabling ships to position themselves within approximately two thousand meters of their location. OMEGA was operational from the beginning of the 1970s up until 1997, when the operation of the Global Positioning System made

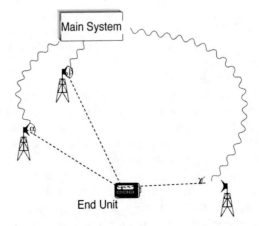

Figure 4.1B Location using AOA technology.

it obsolete. One area in which the service provided by the OMEGA system is still missed is the navigation of submarines in arctic seas under ice. GPS signals do not penetrate the ice; the OMEGA signals, using VLF, were accessible from under the ice (Bowditch 1995).

LORAN (LOng Range Aid to Navigation), or LORAN-C as the current version is called, is an American TDOA navigation system that is still in use. This system, based on "master" and "slave" transmitters invented by Alfred Lee Loomis, was first operated in World War II under the name LORAN-A and consisted of over seventy transmitters covering 30 percent of the world's surface (Bowditch 1995). Technologies that were developed in the late 1940s and the 1950s allowed for the design of a longer range and more accurate service than the service provided by LORAN-A. Still operational today, LORAN-C was launched in 1957 and became the most widely used navigation system—that is, until the introduction of GPS. The system did not cover the whole world but had coverage that was adequate for marine navigation and had an error margin of between 185 and 465 meters.

With the introduction of GPS, fewer funds have been dedicated to the upkeep of LORAN-C, and the system is in danger of being phased out. Advocates for the system argue that maintaining a source of navigation signals other than those transmitted from satellites in case of jamming or failure of the satellite-based systems is essential; in addition, they feel it crucial inasmuch as it functions in places where GPS signals cannot be received.

Tracking stolen vehicles is a lucrative business for some companies, which install tracking devices on vehicles and track them after they have been stolen. Examples of companies that provide a service of this nature are OnStar (www.onstar.com), LoJack (www.lojack.com), and Ituran (www.ituranusa.com). LoJack and Ituran's service are built on land-based location systems.

LoJack Ltd. operates a service that claims to have a 90 percent success rate. The company installs RF transmitters on vehicles that subscribe to the service; it also installs RF receivers in law enforcement vehicles. If a car is reported stolen, the RF receivers installed in law enforcement vehicles automatically attempt to locate the unique signal from the transmitter installed on the stolen car. The transmitter's signal can be picked up within a radius of over two miles.

Ituran Ltd. is another example of a vehicle anti-theft service. Ituran started out as a small company that bought the rights to a technology developed in order to locate military pilots shot down in combat. However, unlike LoJack, Ituran's procured system was based on TDOA technology. Ituran took the technology that it bought and adapted it to serve as a vehicle-tracking system. In order to offer this service, the company builds a local network of antennas and installs transmitters on vehicles that subscribe to the service. When a vehicle is reported stolen, or when the vehicle's alarm system is tampered with, the transmitter begins to emit a signal that is picked up by the system's antennas and the vehicle can then be located. The main advantage of this system over similar systems employing GPS technology is the ability to track vehicles parked in underground parking lots and in other locations in which GPS-based systems do not function. Outside of Israel, where the system is available on a national scale, Ituran offers service in some large metropolitan areas in the United States, Argentina, and Brazil.

Ituran has explored the option of offering its service to track people. Tracking vehicles is technically easier than tracking people because of the ever-present power supply and the ability to install large systems with antennas without causing any inconvenience. Ituran developed a unit that is lightweight and easily carried around (see Chapter 6), but the service was not adopted by enough users to make it profitable.

ANGLE OF ARRIVAL (AOA)

Like TDOA, AOA technology is based on a system of land-based antenna stations. In this case, however, each station, will, as a rule, feature three antennas all pointing in different directions. Once it receives a signal from an end unit, the station calculates the angle from which the signal was sent. It is then passes on this information to the main station, which, based on figures obtained from at least two stations, calculates the point where the two (or more) angles intersect and tracks the end user's location. Unlike the TDOA system, which is dependent on data received from at least three stations in order to determine the end user's location, the AOA system requires data from only two. Figure 4.1B illustrates the process of locating an end unit with an AOA system.

One of the main difficulties that this technology has yet to overcome is differentiating between signals that are received directly and multi-path signals that have bounced off of other objects; these should be ignored since the direction they arrived from has little to do with the location of the end unit. Another difficulty in implementing this kind of system is the need to install several antennas in each base station, leading to costly infrastructure. It is primarily for these reasons that there have been no large AOA systems used for navigation but rather a few small local systems in several places around the world (such as a vehicle location system operated by Pointer Telocation Ltd. in Israel, www.pointer.com).

CELLULAR TRIANGULATION AND
CELL SECTOR IDENTIFICATION

Although the commercial use of cellular communications began as far back as 1983, due to the high price of both the service and the devices, use was limited primarily to business purposes. Cellular phone prices began to drop drastically in the mid-1990s, and today, in the developed world, cell phones are owned by people of all ages, professions, and income levels. Cell phone penetration in the developed world recently crossed the 80 percent mark (Eurostat 2005). In 2005, the United Kingdom had a 102 percent penetration rate, trailing closely behind Israel, with a penetration rate of 106 percent, and Sweden, with a rate of 103 percent (World Bank 2006) see Figure 4.2.

In recent years, the penetration of this form of communication technology has accelerated in many parts of the developing world as well. It is expected that by 2010 more than 50 percent of the world's population will own a cellular phone, at which time we will be able to refer to human society as a whole as a Cellular Society.

Operating a cellular phone network requires that the network operator be able to constantly detect the subscriber's proximity to a specific antenna ("cell"). This enables the operator to transmit incoming and outgoing calls to and from the user's handset. This feature allows for the tracking of the device; however, it is clear that there are more accurate tracking technologies currently available, such as GPS for example (Shoval and Isaacson 2006), which is increasingly being embedded into cell phones nowadays.

The fact that an ever-increasing proportion of human society constantly carries a tracking device at all times and in all places creates new possibilities for spatial research. If all of the phones that belong to the network are tracked at specific periods, cell phone location data can be used for the aggregative analysis of human activity, in practice the Human Sensing of entire populations (Shoval 2007). The application of this aggregative

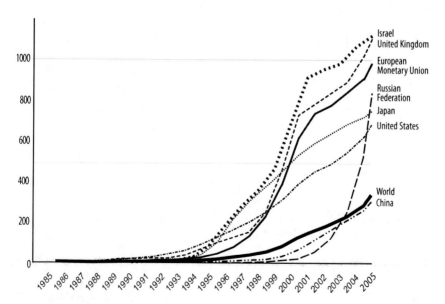

Figure 4.2 World-wide cell phone subscribers (per 1000 people).

approach makes it possible to take a synoptic view of the time-space activity of hundreds of thousands of people in urban and metropolitan areas, or even the time-space activity of millions of people at a national level (Sheridan 2009). This fact, coupled with the considerable progress in the field of GIS, currently places us on the verge of a veritable revolution in human time-space activity research.

Cellular technology has emerged with the tremendous spread of cellular communications networks. Cellular communications networks need to keep track of the location of all of the end units within the system in order to deliver incoming calls to the end units as they are received by the network. Thus, a by-product of cellular networks is actually a location system.

Before proceeding, a few words on how cellular systems operate: Cellular systems get their name from the geographic cells of service that the system is divided into. The size and shape of the cells varies throughout the network and is determined according to the population of subscribers that requires service in a specific location. Urban areas, naturally, have a denser cell layout, leading to smaller cells, while rural areas have cells that cover larger areas in a cellular network. Each cell is equipped with a base station that controls communication throughout that cell. One of the important features of cellular networks is the ability to transfer calls from cell to cell seamlessly without the user feeling that he or she has been passed between cells.

End units can be located throughout the network using cellular cell identification, a process that identifies the cell that the system associates the end unit with, in which the end unit is most probably located. This is only "probable" because, in the event that the system is busy and experiencing a high volume of calls, a unit can be allocated to a nearby cell and not the cell that the end unit is physically in. This is especially true in urban areas where cells are small and base stations are located within close proximity of one another. Using the method of cellular cell identification, the accuracy of the location of the end unit relates directly to the cell size in the specific area. In urban areas with smaller cell size, the ability to locate an end unit using this technique will be more exact, and can even be as accurate as 100 meters or less; in rural areas, cellular cells are large and can be kilometers apart.

Figure 4.3 demonstrates the results of tracking a cellular phone using cellular cell identification. In the experiment, the participant drove with the phone from northern Jerusalem into the Old City of Jerusalem and was located every five minutes. The locations obtained are actually the locations of the antennas of the cellular network that was used and not the result of an average calculated between locations.

Another option, called cellular triangulation, uses the same principles as TDOA, AOA, or monitoring the signal's strength in order to estimate the distance between the end unit and the base station. Once it has estimated the distance, it can triangulate the location of the end unit based on the data

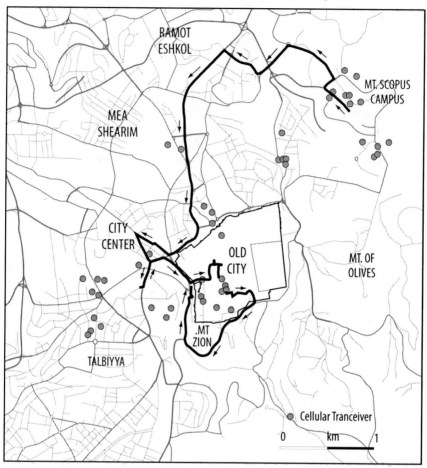

Figure 4.3 Results of tracking a cellular phone using cellular cell identification.

retrieved from three or more base stations. Using this method, the deviation between the position recorded by the system and the end unit's real location can be much smaller than when identifying the cell that the unit is in as described above. Triangulation results can range in accuracy from anywhere between several dozen meters, in ideal conditions, to several thousands of meters, in less than optimal conditions. This method gives more accurate results than the results obtained from cellular cell identification alone, since it is able to reach a resolution below that of the network's cells but similar to the accuracy of cellular cell identification. The variation in this method's accuracy is dependent on the number and density of the network's cells and the volume of activity at any given moment. With cellular cell identification, the more active cells there are on the network, the smaller each cell is, allowing for greater accuracy. Similarly, in triangulation, most cellular networks will prioritize the data transmitted by their antennas. The cells closest to the antenna will be the ones used in the triangulation procedure, particularly during periods of low activity.

RADIO FREQUENCY IDENTIFICATION (RFID)

Radio Frequency Identification systems are based on the broadcasting of electromagnetic signals for short distances. These systems consist of three components: RFID tags, also called transponders; tag readers that are in essence antennas; and a central system that collects the data and calculates the locations of the tags. RFID tags can be either passive tags that contain no battery or power source or active tags that have an independent power source. Passive tags use the electromagnetic current created by the electromagnetic signal emitted from the antenna to power the tag (Spek 2008A, 30).

The advantage of not needing a power source is expressed in the minimal size and weight that tags of this kind can be. These tags can be printed on stickers and embedded into many commercial products without demanding the changing of batteries.

The disadvantage of not having a power source is in the weak signal these tags emit, necessitating that they be close to the antenna in order for their signal to be picked up. Active tags can be picked up from larger distances than can passive tags, but they are larger and need their batteries changed periodically. Tag readers need to be placed throughout the location in which one wants to locate participants and a central system collects the data from the different tag readers and combines them, calculating the location of each tag.

RFID tags have been the center of public discussion focusing on the implications that this technology has on personal privacy. RFID tags have become so small that they can be implanted within one's body. These tags could be broadcasting confidential information that may be picked up by unauthorized sources, leading to an invasion of privacy (Sieberg 2006).

One company that has created an accurate location system by combining RFID technology with TDOA and AOA location abilities is Ubisense Ltd. (www.ubisense.net). The company has developed a system that can locate tagged objects with an accuracy of a few centimeters by tracking the accurate path that the signal traveled before it was picked up by a sensor. In order to use this system (as with all other RFID systems), a network of sensors must be installed.

The requirement of installing sensors throughout the area within which the tracking is desired limits the use of the technology in the context of tourism research to highly defined areas such as museums, shopping malls or self-contained tourist attractions and makes it a costly solution for tracking people over large areas.

BLUETOOTH

Bluetooth is a wireless communication protocol that was developed for communication between mobile and stationary devices over short distances. Many devices that are commercially available have Bluetooth embedded within them. These include printers, PDAs, laptop computers, mobile phones, and many more electronic devices. When Bluetooth is activated in these devices, a radio signal is broadcasted every few seconds. These signals include an identification number, known as a MAC (Media Access Control) address (the process is similar to the address given to a computer when using a wireless network). The device can then be detected by Bluetooth sensors.

Bluetooth sensors placed in known locations can track the number of Bluetooth devices that pass within their range; assuming that these devices represent the number of people who pass by that point, the sensors can sense how busy the area is at a certain time or what the flow of traffic is like. Another use of Bluetooth sensors can be the creation of a database of all of the MAC numbers that have passed within the range of the sensor. Comparing the logs of different sensors located at different locations in the world can demonstrate patterns of migration and movement. If the sensors are located in different countries or even continents, the data collected can reveal patterns of international travel and international tourism. For more information on the information that can be extracted from Bluetooth sensors see www.bluetoothtracking.org.

USING LAND-BASED TRACKING
TECHNOLOGIES IN RESEARCH

Land-based technologies have a great advantage in the fact that they do not need to be directly exposed to the source of the electromagnetic signal or antennas of a system in order to function. This is especially important for tourism research because it allows the tourist to carry an end unit in his or her bag without having to do anything active to participate in the study (Shoval and Isaacson 2007a). The low frequency of the electromagnetic waves that these systems use spreads evenly without the need for a direct line of view between the antenna and the end unit.

RFID technology seems especially promising for tourism research. The idea that one can "tag" a tourist as he or she arrives at a destination and follow the tourist throughout a visit without requiring that he or she carry a heavy device and without the need to power the end unit is especially intriguing, raising privacy concerns that must be addressed and cannot be ignored. The topic of privacy will be discussed in Chapter 9 of this book.

That said, there a few disadvantages that make land-based technologies difficult to implement and not always suitable or possible for use in tourism research. Generally speaking, land-based systems are local systems that must be erected in a specific contained location. Aside from existing systems in a few select places in the world, these systems are not widespread. Even cellular communication networks that, in recent years, have become distributed far more widely, are not available everywhere. RFID systems, which seem promising, are very costly to erect over large areas and are therefore, with the current technological possibilities, not suitable for studying tourism at a municipal level.

Another disadvantage of these systems is that pinpointing locations, in most cases, is accomplished by a central system. This means that as a user of the system, one's location is logged in the system's computers; the system management's cooperation is needed to export these data in order to use them for research. This also means that the main system has to divide its resources among the users of the system, making adding new users and greatly increasing the number of users potentially, costly (Shoval and Isaacson 2006).

5   Satellite-based Tracking Technologies

Since the dawn of civilization, humankind has been inventing ways to answer the simple question: "Where am I?" One of the primary ways this question has been answered, both at sea and on land, has been by gazing up at the celestial configuration above and using the position of the stars as a way to position ourselves down on earth. For thousands of years, humankind has looked up at the sky for guidance and has followed the path illuminated by the stars; the paths of the sky guided our paths upon earth. The launching of the satellites that comprise the Global Positioning System (GPS) has brought the ancient tradition of seeking guidance from above into the digital era.

GPS navigating systems have become an integral part of our lives and are used widely in many commercially available technologies. As a by-product, GPS provides a precise time reference that is used in many technologies that do not have spatial aspects but need a very accurate time reference, for example the synchronization of communication systems such as cellular phone transceivers. GPS technology has become so widespread, in fact, that it is embedded in applications that one would never have expected.

This chapter will introduce the reader to the main satellite navigation systems available today and explain the technology behind these systems. A deeper understanding of GPS technology and its limitations is vital for the greater goal of our book; it can help us exploit all of the advantages of the technology when used to collect data regarding the spatial behavior of tourists.

THE HISTORY OF GPS

Beginning in the early 1960s, the U.S. Department of Defense (DOD) began pursuing the idea of developing a global navigation system. During the 1960s, several projects studied the probability of using electromagnetic signals emitted from satellites for creating a global, continuously available, highly accurate positioning and navigation system.

One of the first efforts was the "transit system," developed in 1965 to meet the navigational needs of submarines carrying Polaris nuclear missiles. These submarines needed to remain hidden and submerged for months at a time, but gyroscope-based navigation, known as inertial navigation, could not sustain its accuracy over such long periods. The U.S. Navy, Air Force, and Army funded and managed projects experimenting with developing technologies and systems that would allow for global navigation. These experimental satellite programs developed many technologies that were later important in developing the Global Positioning System as we know it today and became the building blocks on which today's GPS was developed.

In April 1973, the deputy secretary of defense, Bill Clements, resolved that the Air Force would consolidate the various satellite navigation concepts being developed by the different branches of the U.S. military into a single comprehensive DOD system to be known as the Defense Navigation Satellite System (DNSS). The results of the consolidation of efforts by the different agencies led to the configuration of the GPS as we know it today (Parkinson 1994).

During the 1970s, the first phase of the program was implemented. This phase's aim was to test the feasibility of implementing a full-scale satellite navigation system according to the combined plans. The analysis in this phase included the testing of many new technologies and the introduction of technologies into outer space for the first time. This phase clearly showed that it was possible to use signals transmitted from satellites for navigation; a prototype of the satellites that would be manufactured in the next phase was created.

Between 1978 and 1985 eleven satellites were launched. These satellites, called Block I satellites, were configured to support the system's test program. During this period, the first receivers were manufactured and placed upon a variety of different vehicles—helicopters, airplanes, and cars. In addition, they were carried around by military personnel using twenty-five-pound packs. These receivers allowed for the testing of both the satellite configuration and the function of the receiver design in different conditions and situations.

THE CONFIGURATION OF THE U.S.'S
GLOBAL POSITIONING SYSTEM

The GPS consists of a series of satellites that orbit the earth broadcasting signals, which are in turn picked up by a system of receivers. Although there is a Russian GPS (Glonass) in operation and plans for a future European GPS, the best known and most commonly used of the Global Positioning Systems is the U.S. Department of Defense's "Navigation System with Timing and Ranging" (NAVSTAR). Consisting of twenty-four satellites arranged in six orbital planes, NAVSTAR was, as we have seen, originally conceived as a military navigation system.

Though fully operational since 1994 (Kaplan 1996), the system was initially available to military personnel only, with the DOD deliberately degrading the satellites' civilian signal in order to deny civilians access to its system. In May 2000 the DOD terminated the Selective Availability (SA) procedure, as it was known, opening up the system to individuals and for commercial applications across the globe. The result was that usage of the American GPS became so widespread that the term GPS is, at present, virtually synonymous with the DOD's NAVSTAR system (Getting 1993).

The GPS contains three separate operational segments:

1. The Space Segment (SS), consisting of the satellites that orbit the earth;
2. The Control Segment (CS), consisting of the three different kinds of control centers located on the ground; and
3. The User Segment (US), made up of all of the receivers that are used by different people to acquire positions.

The Space Segment

This segment consists of the satellites that orbit earth and emit the electromagnetic signals that are then used to obtain locations. The satellites orbit earth at a height of 20,200 kilometers in nearly circular orbits. Originally, this segment was planned to include twenty-four satellites on six orbital planes. Twenty-one satellites are essential in assuring that the minimum of four satellites needed for navigation are within a direct line of sight at all times from any location on earth. Three more satellites were included in the original satellite configuration in order to serve as redundancy in case of the failure of any of the twenty-one essential satellites. At the time of writing, there are thirty-one operational NAVSTAR GPS satellites. The additional satellites add to the reliability and accuracy of the system and allow for the maintenance of the system with minimal loss of navigation abilities.

The satellites' main goal is to orbit earth emitting navigation signals that, when picked up by a receiver, allow the receiver to calculate their locations. Onboard the satellites are atomic clocks that ensure that the timing aspect of the signals is accurately synchronized. These atomic clocks and the central importance of accurate time make the GPS signals a good tool for calibrating different systems that rely on accurate timing (for example, communication systems that use time-based technologies), allowing the achievement of high accuracy without the purchase of costly atomic clocks.

The Control Segment

This segment consists of the control stations located on the ground. These stations control the satellites from the ground and have two main functions. The first role is to maintain the satellites' orbit pattern and to keep each satellite in its designated position. The second role is to synchronize the atomic clocks on board the satellites so that all of the clocks are within only a few nanoseconds of each other. If any of the systems on board the satellites fail, the ground stations help in fixing the failing systems and compensating, using other satellites so that the navigation abilities are only minimally affected.

Three types of stations make up the control segment:

1. Master Control Station (MCS): One MCS, originally located in California at the Vandenberg Air Force Base, was later moved to the Consolidated Space Operations Center (CSOC) located within the Shriver Air Force Base, Colorado Springs. The MCS is responsible for the overall function of the NAVSTAR GPS system. In order to maintain the system, the CSOC receives data collected in the monitor stations and calculates the satellites' orbit and clock parameters using a Kalman estimator. Kalman estimators make it possible to assess the state of a dynamic system using a series of inaccurate and incomplete readings (El-Rabbany 2006). The calculations of the satellites' orbit and the clock parameters are done in real time so that any malfunction of the system can be detected immediately.

2. Monitor Stations: Six monitor stations spread out around the globe belong to the official network that is used to calculate the ephemerides, the location of the satellites in the sky at a given time. The Monitor Stations measure the range to every satellite in view every 1.5 seconds and transmit the data collected to the MCS every fifteen minutes. Apart from the official GPS control segment described above, several other networks measure accurate orbital information. These networks include, among others, the Cooperative International GPS Network operated by the U.S. National Geodetic Survey and the International GPS Service for Geodynamics operated by the International Association of Geodesy. These alternative networks collect and distribute raw tracking data and the satellite clock parameters that are used for geodynamic applications, which require the highest accuracy possible.

3. Ground Control Stations: Four of the Monitor Stations serve as control stations. These stations have the equipment needed to broadcast data up to the satellites. The satellites' ephemerides and clock synchronization information are received from the MCS. This information, crucial to the functioning of the system, is broadcasted up to the satellites once or twice a day.

The User Segment

This segment consists of all of the receivers that serve end-applications in interpreting the GPS signals and use them for navigation. No matter how small—and in recent years there has been great progress in resizing GPS receivers so that they can even fit into wristwatches and other small devices—all receivers are made up of an antenna and a receiver-processor that calculates the locations based on the signals that are received. Receivers can also include displays and buttons that are used by the user to display different parameters of movement or navigation. Within the User Segment, GPS is a one-way broadcasting system. The satellites send a signal, which is then picked up by essentially "passive" receivers. This means that, much like television or radio broadcasting systems, the system can support an almost unlimited number of end users (Zhao 1997).

OBTAINING LOCATION USING THE GPS SIGNAL

A GPS satellite transmits a continuous signal that contains the precise time of day as received from the atomic clock, the ephemeris data, and the almanac data. The receivers use the information about the time of day to calculate the distance between the receiver and each of the satellites whose signal is obtained. Distance is calculated by the difference in time stamps between the time when the signal was sent and the time when it was received. Electromagnetic signals travel at the speed of light so that the distance traveled can be calculated by multiplying the time that it took for the signal to reach the receiver and the speed of light.

The signal from the first satellite creates a sphere around the satellite at the distance calculated (see Figure 5.1). Adding the distance calculated from the second satellite creates an additional sphere. The intersection of the two spheres results in a ring. Intersecting the previous spheres with the sphere around the third satellite results in two points. The receiver can now eliminate one of the points as impossible due to its distance from earth and can calculate the longitude and latitude but not the exact height. Navigating this way, using only three satellites, produces two-dimensional locations. In order to calculate location in three dimensions (longitude, latitude, and height) the sphere calculated around a fourth satellite is needed. When more than four satellites are visible, the residual information from the satellites is used to improve the accuracy and to estimate the error.

POTENTIAL SOURCES OF ERROR IN CALCULATING POSITIONS

The accuracy of Global Positing Systems varies greatly and is dependent on the nature of the terrain (open rural as opposed to dense urban), weather conditions, and the extent of the GPS receiver's exposure to the sky. The

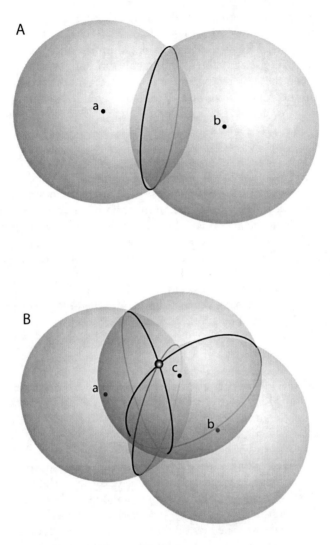

Figure 5.1 Three-dimensional triangulation.

receiver will provide an accurate reading only if directly exposed to the satellites' signals. Any obstruction, regardless of whether it partially or wholly blocks the signal, can result in an imprecise reading. Many different kinds of obstructions exist; we will mention and explain the main sources.

Atmospheric Interferences

The not-so-uniform nature of the atmosphere is a source of error in the distances calculated by GPS receivers. Humidity, changing conditions in the ionosphere, and different altitudes of satellite locations all influence the speed at which the satellites' signals travel, slowing the signals down and causing a delay in the time when the signal is received. The extent to which atmospheric conditions affect the signals is not uniform and is a result of the location of the different satellites. The signal from satellites that are directly overhead travels the shortest distance and is therefore the least affected by the conditions in the atmosphere. The signal from satellites that are close to the horizon travels for a longer distance and therefore is affected the most.

Multi-path Errors

Another source of error exists in dense urban areas, referred to as "urban canyons." In dense areas the signal bounces off of hard surfaces such as buildings and sidewalks, creating an indirect path between the receiver and the satellite. When the signal is deflected off of these objects on its way to the receiver, the signal received has been delayed and the distance calculations include an error of the multi-path that the signal had traveled. These "bouncing" signals, together with the obstructed view of the sky, make dense urban areas more challenging for the function of GPS. The longer the distance traveled by the signal due to bouncing, the greater the error in the position calculating will be.

Indirect Exposure

GPS receivers need direct exposure to the sky in order to operate optimally; any obstruction of the signal transmitted from the satellites can cause a degradation in the accuracy of the location obtained and can lead to the total inability to calculate the location of the receiver (Xu 2007). The most common obstruction encountered is the roof of buildings. When located indoors or in a covered area, the GPS receiver will have a hard time obtaining an accurate location. Depending on its sensitivity, the receiver might obtain low fortitude signals that penetrate windows or the roof of the building (depending on the roof's material). This results in poor accuracy in the location calculated or in the loss of ability to calculate locations at all.

GPS tracking of tourist activity faces similar challenges. One of the main issues is the constant combination of indoor and outdoor environments that tourists walk in and out of. The GPS's main advantage as a tool for tracking pedestrian activity lies in the fact that it is a global system. Its main disadvantage is that it requires, at all times, a direct line of vision between the receiver and the sky. This makes tracking in dense urban locales difficult, limiting the areas in which the system can provide an accurate reading to open venues only.

Time Passed from Last Ephemeris Calculation

The ephemeris data, including accurate data about the location of the satellites, are transmitted every thirty seconds but are only calculated every two hours. This information is critical to the accuracy of the locations calculated. Any deviation of the satellite from the known location results in an error in the location calculations. As time passes from the last ephemeris calculation, the accuracy of the location of the satellites lessens, resulting in a degrading in the accuracy of the locations calculated.

Accurate readings of time are essential for the accurate calculation of locations. While the satellites have atomic clocks on board, most receivers (excluding very expensive ones) do not have atomic clocks. Atomic clocks are very costly and still relatively big and therefore installing them within receivers would make the whole technology too costly to implement for most purposes. The receivers tune their clocks according to the signals that they receive from the satellites and are therefore very accurate, but even a small inaccuracy of a few milliseconds can result in a few meters of inaccuracy.

Intentional Degradation of the GPS Signal

Intentional degradation of the GPS signal can also be a cause of error in the locations obtained from a GPS receiver. As mentioned above, the GPS signal was intentionally degraded for civilian users up until May 2000 when Bill Clinton, president of the United States, canceled Selective Availability (SA). Although there are rumors that SA has been reactivated during times of military conflict, the U.S. government denies that this has happened since 2000. In September 2007, President George W. Bush accepted the recommendation of the Department of Defense to end procurement of GPS satellites that have the capability of intentionally degrading the accuracy of civil signals. This decision did not improve the quality of navigation throughout the world but rather clarified the intentions of the U.S. government regarding the future of the GPS (The White House 2007).

IMPROVING GPS ACCURACY

Of the several technologies that have sought to increase the accuracy of GPS readings, two of them, Differential GPS (DGPS) and Wide Area Augmentation System (WAAS), have been successfully implemented. The idea that is at the base of both technologies is the same. Both technologies' goal is to find a way to correct the error in the signal that is received from the satellites. The source of error in the signal originates both in various interferences that the signal encounters on its way from the satellite to the receiver (such as atmospheric interferences or multi-path errors, discussed above) and in the inaccuracies of ephemeris data that the receiver uses. Both technologies use stationary receivers. Being located at known locations, these receivers are able to calculate the error in the signal received from each specific satellite and then instruct the roaming receiver as to correcting the signals obtained.

DIFFERENTIAL GPS

DGPS assumes that receivers that are located close to each other will experience similar atmospheric and terrain interferences. The DGPS uses a stationary receiver in order to correct the signal obtained by a roaming receiver. It does so by comparing the accuracy of the positions each receiver obtains from the system's satellites with the signal of the stationary receiver. The stationary receiver, called the reference station, can either broadcast a correction to the signal received from the satellites or save the correction for it to be combined with the data that the receiver obtains at a later time. In order to utilize the calculated correction to the maximum effect, it is important that the reference station and the roaming receiver be located close to one another.

Two possibilities exist for the implementation of DGPS: The first is to set up a reference station at a location that is close to the area in which the locations are collected and that has been surveyed very accurately; the other is to receive corrections from a reference station that was erected by someone else but that transmits the information collected to the public. While less expensive and easier to implement, this second option is not always possible to use, as not every location has an available DGPS service. In order to be effective, the reference station must be located in close proximity to the receiver, ensuring that they have the same physical conditions. Using DGPS can improve the accuracy of the location calculated to an accuracy of less than one meter.

WIDE AREA AUGMENTATION SYSTEM

WAAS is a North America-based air navigation aid system (Federal Aviation Administration 2007). The system is designed to provide the additional accuracy, availability, and integrity that are needed when navigating an aircraft. The WAAS consists of twenty-five ground reference stations. The stations, which are distributed across the United States, help when calculating an end user's position by assisting to factor in the signal delay caused by various impediments, as discussed above. The information gathered by the reference stations is forwarded to two central stations, which, by correlating all of the incoming data, obtain a correction to the signal received from the satellites. The adjusted differential message is then broadcast from two geostationary satellites and picked up by the roaming receivers.

Although planned to assist in aerial navigation, WAAS is effective in achieving greater accuracy in GPS locations for other receivers as well and can improve accuracy to between two and three meters. WAAS is not as accurate as DGPS due to the large distance—and therefore different conditions—that may exist between the ground reference stations and roaming receiver. Owing to its sizeable and complex infrastructure,

WAAS is currently only available in the United States and in small parts of Canada and Mexico.

OTHER GLOBAL POSITIONING SYSTEMS

The popularity of the Department of Defense's GPS and its widespread use have made other countries want to develop their own systems. The idea that so many technologies rely on the American system, which can, theoretically, be disabled at any time by the U.S. military has caused great concern to other nations. The Russian GPS known as GLONASS is the second most developed GPS available today. GLONASS was developed by the Russian Aviation and Space Agency (currently the Russian Federal Space Agency, RSA). GLONASS, like the American GPS, was designed to have twenty-four satellites but, unlike the American GPS, the GLONASS satellites are arranged in three orbital planes. As of the end of December 2007 only thirteen satellites were operational, giving partial service to all parts of the world. They are deployed in a way that gives preference to achieving navigating abilities in specific areas such as the former Soviet Union. GLONASS has a closed military segment whose signal is only accessible for Russian military, but it also has a segment of communication open for civilian use. On the market today one can purchase receivers that navigate using the GLONASS signal.

In 2004, Russia and India announced their plans to cooperate on the GLONASS project. Indian territory was supposed to be covered by the system and the Indian army was supposed to be able to access the coded signals in exchange for the Indians covering some of the development cost and launching satellites using their facilities and technology. These plans for cooperation have yet to become a reality and, at the time of writing, not much has been done to further this initiative.

The European Union has made a decision to invest 3.4 billion Euro in its GPS, called "Galileo." The decision from November 2007 (BBC News 2007) included a deadline for the system to be operational by 2013. This initiative is different than the DOD's GPS and GLONASS due to the multinational nature of the EU. The EU has involved private consortiums in the development of the project and later these companies will take an active role in the day-to-day management and upkeep of the system. The EU sees this project as a way to ensure their independence and guarantee that they will not need to rely on American technology (Lembke 2003).

HYBRID SOLUTIONS

This technology combines several geo-location technologies, seeking to reap the benefits of each while minimizing their various disadvantages. Of the hybrid solutions available today the leading one is Assisted GPS (AGPS). This technology uses GPS technology in combination with a land-based antenna network in order to pinpoint specific locations. It was originally conceived as a means of locating the position of mobile phones within a cellular network with greater accuracy than was possible when using cellular triangulation alone. In this method, the land-based stations are equipped with GPS units, which are used to predict the signals picked up by the roaming receivers. This means that end units can be fitted with only a partial, hence much smaller, GPS receiver.

This amalgamation of GPS and land-based networks has several advantages. It provides much more precise readings indoors. It also solves the problem of having to incorporate unwieldy GPS receivers into today's trendy, miniature handsets (Djuknic and Richton 2001).

GPS RECEIVERS ON THE MARKET

There are many different GPS receivers on the market today. Over the past decade, GPS receivers have become more sophisticated, include more features, and are much less expensive than they used to be. Different receivers are manufactured and marketed with features that aim to serve different kinds of GPS users. The features that are incorporated into the receiver, along with the size of the receiver and the additional electronics and software, determine the price of the whole package. Consumers searching for a receiver must take into consideration a number of features that may differ from model to model.

The first receivers on the market were designed with a single channel with which they could lock onto the satellite's signal. This meant that they could only lock onto one satellite at a time. It would take these receivers a relatively long amount of time to achieve a lock on four satellites and enable three-dimensional navigation. Most of today's receivers are equipped with a twelve-channel design, allowing them to lock on to twelve satellites at once and thus greatly reduce the time that it takes to achieve an initial position.

Another feature that can vary greatly between different GPS receivers is the battery life. Great advances have been made in the technologies that determine the size of power cells; the batteries that are now available are much smaller and more powerful than those used ten years ago. That said, there is still a very big difference in the battery life of different receivers and it is important to ensure that the battery will be sufficient for the planned use.

Another feature that may differ is the receiver's support of accuracy-enhancing technologies such as DGPS or WAAS. These features may be helpful in achieving greater accuracy but are not suitable for use in every research project—for example, projects not located in North America cannot use WAAS. Projects that track the routes people travel need receivers that are able to save information about an entire route or track. This means that the receivers must have both memory space for saving the track and software that enables the creation of a log.

Surveyors use GPS to make high-accuracy measurements. They are willing to pay a high price, both in money and in comfort, for equipment that will give them accurate locations. Receivers manufactured for this market can be very large, and may include large external antennas and large computers that run designated software including specialty features that enhance accuracy as well as DGPS and WAAS capabilities. These receivers can be very expensive and may cost anywhere from $1,500 to well over $10,000.

Another type of specialty GPS receiver is manufactured for use within airplanes or on board ships. These receivers include special features that help navigate in the special environment within which they are used.

GPS receivers that are manufactured with the recreational user in mind have become quite diverse in their uses. The first recreational receivers available were designed to be used together with a map. Their main use was to calculate the coordinates of their location and then present them on a screen. The user would then look the coordinates up on a map and see where he or she was located. Additional features included entering coordinates manually and then navigating toward a location or committing the coordinates to the device's memory.

There are receivers that are incorporated in a wristwatch and include physical training software. In these watches, the information collected using the GPS receiver helps in calculating the distance a person walked or ran as well as the speed and other parameters of the workout.

Another widespread use for GPS is as a navigation aid when paired with navigation software. For this use, a whole market of dedicated hand-held computers (PDAs) that have integrated GPS chips and run navigation software exists.

Another configuration that can be used to navigate is using a Bluetooth-enabled GPS receiver together with a PDA that supports Bluetooth. Using this configuration it is possible to position the receiver in a place that has better satellite reception than the place where the PDA is placed. This configuration also allows for the use of more than one kind of GPS receiver (though not at the same time).

Notable types of receivers that have become available on the market in the past year or two are receivers that incorporate a data logger within them (see Figure 5.2 for examples of three receivers with data loggers produced by different manufacturers). These receivers, which may include Bluetooth communication as well, have been developed using the advance in flash memory technologies that allows for the storage of large amounts of data in a small chip. These receivers' abilities to store logs of tracks taken are especially useful for tracking people given that all of the needed functionality (obtaining locations and logging them) is available in one small device. Recreational receivers can be bought at prices ranging from less than $100 up to several hundreds of dollars.

TOURIST-DESIGNED GPS APPLICATIONS

Tourists by definition need to find their way in places that they are not familiar with and, therefore, GPS technology has always seemed to have a lot of potential when it comes to the tourism market. Applications geared at tourists have gone far beyond just helping tourists find their way in unknown locations. They now use technologies called Location Aware Technologies or, as they are better known, Location-based Services (LBS). These technologies, in addition to assisting in navigation, supply the tourist with information that is derived from his or her spatial context (Brown and Chalmers 2003; Schilling et al. 2005). Examples of this kind of service include supplying the tourist with information on the hours of operation or entrance fees for the attractions located nearby, information on the history of the buildings that he or she passes, and information on restaurants or hotels located in close proximity to the tourist's current location.

Another type of service offered is a computerized tour guide that uses GPS technology to locate the tourist and play explanations that are relevant to the location of the tourist at a given moment.

The technology needed to produce GPS-enabled tour guides has been around for a few years already but, despite large amounts of funding that were put towards research and quite a few projects that reached the prototype stage—for example, "M-toGuide" supported by the EU fifth

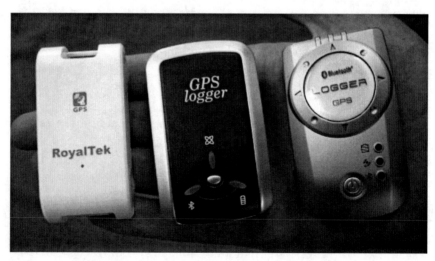

Figure 5.2 GPS loggers.

framework and "Dynamic mobile-assisted Tour Guide" (DTG) supported by the BMWi (German Federal Ministry of Economics and Technology)—the involvement of big corporations (Motorola in the case of M-toGuide and Microsoft in the case of DTG) has not led to the production of large scale commercial GPS-guided tour services.

A new effort that at the time of writing (March 2009) looks promising is the incorporation of a GPS-enabled guided tour format into the new Garmin (one of the largest GPS receivers manufacturers) automotive GPS product lines. This format, known as the "Garmin TourGuide format" allows users to download GPS tours created by a third party onto their GPS units. This project is part of a larger effort that Garmin calls "Points of Interest" (POI), which allows users to download the locations of different points of interest (such as gas stations, rest areas, or other types of services) onto their GPS units.

Part III

Application of Tracking Technologies to Research on Tourist Mobility

6   Methodological Challenges

This chapter focuses on two fundamental research design questions that are relevant to any tourism study hoping to implement advanced tracking technologies. The questions are: (1) Which tracking method is the most suitable for a specific planned research project, depending on the spatial and temporal scales of the study and (2) What are the different possible locations for distribution and collection of the tracking devices. The process of answering these questions in a systematic way will illuminate other issues related to the implementation of tracking technologies for tourism research.

In order to address these questions, this chapter is divided into two sections. The first section engages with the first question, presenting the results of a series of four experiments using the various technologies described in the previous two chapters. These experiments were conducted in tourist destinations of various scales, ranging from tourist attractions to the national scale. The second section in the chapter presents the authors' view of possible research strategies, focusing on tracking methods and sampling strategies for different types of tourism destinations and for specific tourism segments.

THE SUITABILITY OF DIFFERENT TRACKING
TECHNOLOGIES FOR TOURISM RESEARCH

The first section of this chapter details the processes and results of four experiments designed to determine the potential worth of the various tracking technologies discussed in the previous chapters as tools for research into tourist mobility. Carried out in several geographical locations and on different scales, the aim of each study was to establish the advantages and limitations of a specific technology. Aside of the important issue of accuracy, the tests were also concerned with other problems a researcher might face, including the size of the device used and the degree of cooperation demanded from the subject. As noted, if they are to be used for research purposes, tracking technologies must be both small and passive, so as not to disrupt or affect the subject's behavior.

The first experiment, which tracked one subject through the historic area of Jaffa (Israel), compared the results obtained by GPS and a TDOA tracking system. The second experiment evaluated a land-based TDOA tracking system on a variety of geographical scales in Israel. The third experiment took place in Jerusalem's Old City, and compared results obtained using cellular triangulation of a portable phone to those attained using two different types of GPS devices. The fourth experiment evaluated four types of GPS receivers in the PortAventura theme park, located roughly 100 kilometers south of Barcelona.

Experiment I Land-based TDOA Tracking vs. GPS

The aim of the experiment was to test whether modern tracking technologies can be usefully applied to the study of tourists' spatial, and, more specifically, pedestrian spatial behavior. The experiment put to the test and compared the performance of two of the tracking technologies described in Chapters 4 and 5: the GPS and TDOA land-based system. They were tested under identical circumstances and in a variety of situations.

The experiment was conducted in Jaffa's historic center, part of the city of Tel Aviv-Jaffa. Old Jaffa is a mixture of assorted urban textures and geographical locations including a waterfront, alleys, open and covered markets, and an observation point in the middle of a park, which is also the highest spot in the area. This diverse range of urban textures made it possible to examine how the two technologies performed against a backdrop of various physical constraints.

GPS Tracking System

The experiment used a Magellan GPS receiver with an external antenna. The receiver was placed inside a knapsack, while the antenna was attached to the knapsack's exterior. The receiver was set to record a tracking point every ten meters.

TDOA Land-based Tracking System

The land-based tracking technology used in the experiment was a privately run system by Ituran Ltd., which manages a tracking system based on TDOA technology (see chapter 4). Ituran owns and services a TDOA infrastructure offering tracking services in the United States, Brazil, Argentina, and Israel. The experiment used a small hand-held tracking device (Figure 6.1), which can be simply placed in a bag and carried around. The system was programmed to obtain a reading every tenseconds.

Figure 6.1 Ituran TDOA end unit.

During the experiment, the same person carried both units simultaneously. This meant that, all other conditions being equal, it was possible to compare the results of both technologies objectively. During the experiment, the subject walked along a 2.35-kilometer-long circular route for 2 hours and 45 minutes.

Results

Figure 6.2A documents the track obtained by the TDOA system. As noted, the system registered a reading every ten seconds, hence the large number of points recorded. As there are no transceiver stations in the sea, it proved particularly difficult to obtain an accurate reading along the town's coastline, making it very challenging to triangulate the subject's position when in this area. When the system's reading was skewed, it logged the subject's as having moved an unreasonable distance to an impossible location, e.g., the middle of the sea. While regrettable, this also means that researchers can, plausibly, clean up the results obtained by expunging physically unfeasible points.

Figure 6.2A reveals and underlines several of the problems that may emerge when using TDOA technology to track human spatial behavior.

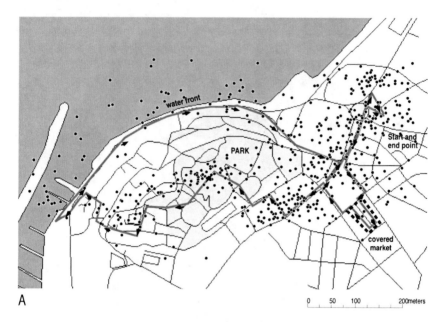

Figure 6.2A Results of tracking in Old Jaffa, TDOA.

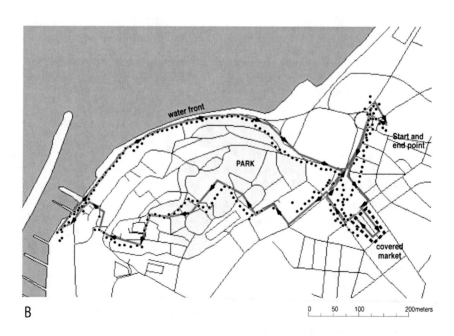

Figure 6.2B Results of tracking in Old Jaffa, GPS.

The TDOA system used in this study implements algorithms that assist in the location of a moving vehicle, algorithms that were developed to enable to tracking of stolen vehicles. Figure 6.3 shows a track created by a car driving along the roads connecting Jerusalem to Tel Aviv; the track obtained is clear and accurate, revealing the movement along the road. Compared to Figure 6.2A, the results obtained when tracking a moving vehicle are far superior to those received when tracking a pedestrian due to the algorithms discussed above. This deficiency was most apparent when the subject was stationary. It is possible, however, that the application of similar algorithms aimed at improving the accuracy of locating pedestrians can substantially increase the system's levels of precision.

Figure 6.2B shows the track obtained from the GPS receiver. It is much tidier than the TDOA track; this is simply because it used a different criterion for logging points—distance as opposed to time. The covered market proved the most difficult area to log. In it, the GPS receiver provided readings that were at times up to fifty meters off base.

The accuracy of a tracking system is an important parameter for the evaluation of the system; this value varies under different physical settings and with the tracking of different objects (for instance, moving automobiles vs. pedestrians, as discussed above). A comparison of results obtained by each system is shown in Figure 6.4. The accuracy of each system can be determined by measuring the distance from the known path to the point located by the system; this measurement is the deviation. Figure 6.4A shows the frequency of deviations in both systems. The accuracy of a system can be defined as the distance from the known path based on a percentage of deviations. For example: Figure 6.4A shows that the GPS system can give an accuracy of up to twenty meters 90 percent of the time. An examination

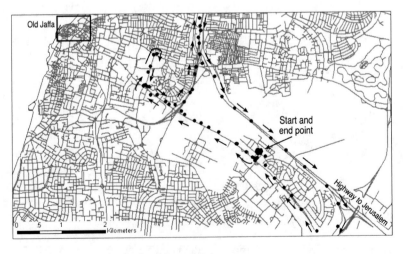

Figure 6.3 TDOA track obtained by a driving car.

of Figure 6.4B reveals that the amount of accumulated deviation decreases over distance. The TDOA system's rate of decrease is a linear decline with a moderate slope. The GPS system's decrease rate is much more rapid; the number of deviations decreases rapidly to no deviations at all. This rapid decrease is illustrated in Figure 6.4B, demonstrating that the GPS system has an accuracy of 90 percent at approximately twenty meters while the TDOA system achieves this accuracy only at sixty meters.

An important feature to note is that the frequency function of the deviations in both systems has a clear tendency to converge. This tendency is what allows us to describe each system's accuracy. Had the frequency function been a divergent function, it would not be possible to estimate the accuracy of the system. The different rates at which each of the deviation functions converges dictates the scale of the research in which they can successfully be used.

Thus TDOA systems can be used successfully in research projects that do not need the greater accuracy that GPS systems can supply, but that can benefit from the ease of use and the extensive coverage that the systems give. On the other hand, research projects designed for a smaller geographic scale that need high data resolution will benefit from the abilities that a GPS system can provide.

EXPERIMENT II LAND-BASED TDOA TRACKING: DIFFERENT GEOGRAPHICAL SCALES

This experiment was designed to determine the value of using TDOA tracking systems to study the spatial behavior of tourists on a variety of geographical scales. As in the previously described experiment, the TDOA

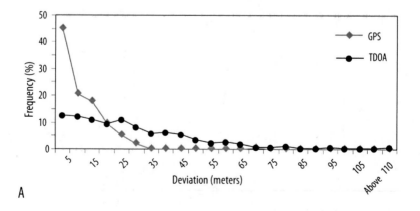

A

Figure 6.4A Frequency of deviations in the comparison between tracks obtained from TDOA and GPS systems.

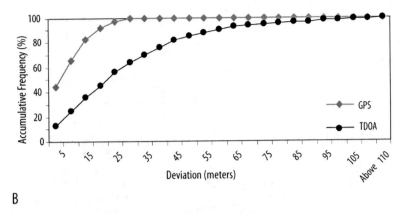

B

Figure 6.4B Cumulative frequency of deviations in the comparison between tracks obtained from TDOA and GPS systems.

technology used in the experiment belonged to Ituran Ltd. Over the course of the experiment, the subject carried around a small tracking device, which in this instance was placed in a knapsack. The system was programmed to record its location once every minute.

The experiment took place on June 2 and 3, 2005. The device was carried by a student while on a field trip to Nazareth and Akko (Acre) in the north of Israel. Akko is most famously known for its underground Crusader City, and in 2002 UNESCO added the town's Old City to its list of World Heritage Sites. The town of Nazareth was first mentioned in the New Testament as the town in which St. Joseph and the Virgin Mary lived. It was in Nazareth that the Annunciation took place and that Jesus spent his childhood and adolescence.

National and Regional Analysis Using a TDOA System

The field trip began near the town of Yokne'am. From there the group traveled to Akko, where they spent most of the day. In the afternoon, they drove to Nazareth, where they stayed overnight in a hotel. The group spent the following day touring the city, after which they traveled back by bus to Jerusalem along Highway no. 6.

As can be seen in Figure 6.5A, which was produced on a national scale using data obtained from the TDOA system, the group spent most of its time in the towns of Nazareth and Akko. The roads used to travel to and from the two towns are also clearly marked on the track. Figure 6.5B, drawn using the same data, focuses solely on the north of Israel. Offering a more detailed picture, this chart was drafted on a regional scale and shows which parts of Akko and Nazareth the group visited and which routes it took to and from the two towns.

Table 6.1 Advantages and Disadvantages of Different Tracking Technologies for the Study of Pedestrian Spatial Behavior

	GPS Tracking Technology	*Land-based Tracking Technologies*
Advantages		
	1. Worldwide availability.	1. Virtually unaffected by weather conditions or terrain.
	2. Due to the system's multitude of satellites, it does not suffer from "downtime" and is readily available at all times.	2. As there is no need to expose the end units directly to the RF stations' transceivers, they can be carried in a bag or placed in a pocket.
	3. While the cost of GPS receivers has fallen steadily I over the past few years, their quality and level of sophistication has increased radically.	3. Work well in dense urban areas as well as indoors.
	4. In optimal conditions even inexpensive GPS receivers can provide an accurate reading of up to a mere few meters.	4. The end units retain power for several days.
Disadvantages		
	1. The receiver cannot obtain a signal if it is located under significant coverage; make it difficult to pick up the satellites' signal.	1. Can be used only in areas that have the appropriate infrastructure, which is privately owned and relatively expensive.
	2. High power consumption by the receivers makes it difficult to track a subject over a long period of time without changing batteries.	2. Each system has its own protocols and requires the use of specially designated equipment.
	3. In dense urban areas structures blocking the sky obstruct the line of sight, diminishing the system's ability to pinpoint the receiver's location.	3. When locating a unit's position, the entire system is activated. Frequent tracking uses up "expensive" system resources and may interfere with the system's ability to provide high-quality tracking for all users.
	4. The current generation of GPS receivers cannot provide information on spatial behavior in indoor environments.	

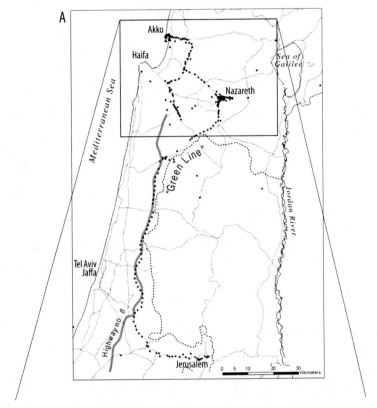

Figure 6.5A Track obtained using a land-based TDOA system—National scale.

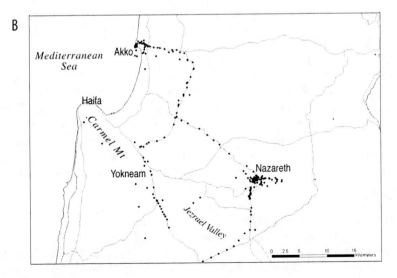

Figure 6.5B Track obtained using a land-based TDOA system—Regional scale.

Intra-City Analysis Using a TDOA System

The path the field trip took through Akko's Old City was as follows (Figure 6.6): The group began the tour at the visitors' center and visited the adjacent archaeological site. It then explored the Templars' Tunnel, having first passed through the local market (suq). This done, the group stopped for lunch at a restaurant overlooking the marina. Walking once again through suq, the group returned to its bus, which was parked near the entrance to the visitors' center.

As can be seen in Figure 6.6, the TDOA system was for the most part unable to track the route followed by the group due to its relatively low level of accuracy, and is thus clearly unsuited for micro-level research. If it is to further the understanding of the spatial and temporal behavior of tourists in small towns and sites, the accuracy of the data obtained at this level must be greater than that obtained at a regional or national level. That said, the TDOA system did indicate, albeit somewhat loosely, the group's location and, as such, can be used for researching tourist activity in larger cities, or in those cases in which the exact position of the tourists is of less importance.

Experiment III Comparison Between Cellular Triangulation, GPS, and AGPS

This experiment compared three kinds of tracking technologies: cellular triangulation, GPS, and the hybrid AGPS. The experiment was conducted in the Old City of Jerusalem. One of the world's oldest continuously inhabited cities, the Old City is full of narrow, twisting streets and alleyways, some only a few meters wide. In addition, many of the Old City's streets are wholly or partially covered. Hence, the Old City's jungle-like density makes it ideally suited to assessing the performance of different tracking technologies when faced with acute physical constraints.

The experiment took place on June 9, 2005. Over the course of the experiment, a research assistant traveled around the Old City carrying, at one and the same time, all three of the aforementioned tracking devices. The 3.1-kilometer track took 1 hour and 26 minutes to complete (Figure 6.7). Beginning at the Jaffa Gate, the research assistant entered the Old City. He then continued, on foot, along David Street towards the Church of the Holy Sepulcher. Crossing part of the Roman Cardo into the Jewish Quarter, he walked towards the Western Wall Esplanade. Once there, he hailed a taxi and returned to Jaffa Gate via Mount Zion.

The GPS tracking system used in this experiment was an Emtac CruxII BlueTooth GPS receiver and Pocket PC (see Figure 6.8). The receiver was set to record its location once every second.

The AGPS tracking system embedded in a cellular phone used in this experiment was a Motorola i860 AGPS-enabled cellular phone linked to the

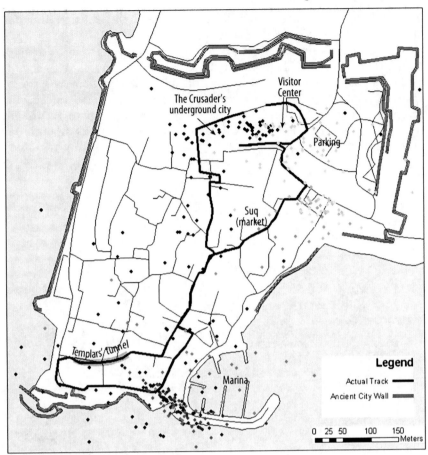

Figure 6.6 TDOA track in Old Akko.

MIRS Communications' cellular network (MIRS being a subsidiary company of Motorola, Israel). The cellular phone unit was placed in the breast pocket of the student's shirt, giving it an uninterrupted line of sight to the sky. The MIRS network was instructed to ascertain the device's location once every thirty seconds. Having done so, it logged the device's position on the network's central server, from which the data were eventually retrieved.

For the cellular triangulation in this experiment, a Samsung 854 cellular phone linked to the Partner Communications cellular network was used, employing an Internet-based graphical user interface. This particular apparatus was designed specifically to help companies' locate their workers during the day. It proved impossible to change the device's fifteen-minute location default frequency, and so a manual location check was done once every five minutes. The system logged the device's position on the Partner network's central server, which was then accessed through the Internet.

The tracking points thus accumulated were eventually transferred manually to a GIS layer. The cell phone was placed in a bag.

The results obtained using the GPS receiver proved accurate at most points (Figure 6.7B). There was some difficulty in pinpointing the GPS receiver's position when it was lodged under the Old City's covered streets. However, once it had left that area, the receiver was able to, once again, promptly fix its location. Given that the system was set to record the receiver's location once every second, the result was a fairly smooth and accurate track.

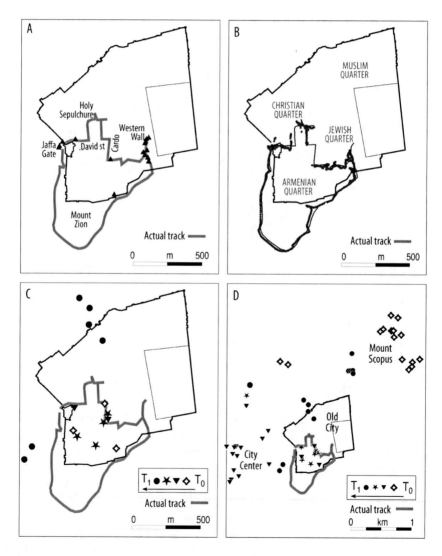

Figure 6.7 Tracks obtained using AGPS, GPS, and cellular triangulation.

The tracking results obtained using the AGPS-enabled cellular phone proved rather disappointing (Figure 6.7A). Unlike the GPS receiver, which was set to retrieve its location once every second, the AGPS unit, owing to the limitations imposed by the cellular network, had a much lower sampling rate. Set to register its position once every 30 seconds, the result included considerably fewer location points. Nevertheless, the points the system did obtain were, on the whole, accurate. It is worth noting that the device found it difficult to log its location during the taxi ride back from the Western Wall area to Jaffa Gate. This was somewhat surprising as it was thought that the system would have no problem in fixing its position while in a car. It is entirely possible that this was the result of the specific AGPS unit being less sensitive than the GPS device used in the experiment and therefore incapable of picking up satellite signals through the roof of the car.

The track obtained using cellular triangulation was, as expected, the least accurate of all (Figure 6.7C). Indeed, the location points obtained were not only few and far apart, but ill defined, sufficient only to note whether the device was near or in the Old City. It thus proved impossible to reconstruct the path taken by the research assistant. Nevertheless, cellular triangulation should not be dismissed as a research tool. As a means for collecting spatial data, cellular phones have many advantages, including and above all the fact that they are also fully functional phones. This is significant; tourists are quite happy to carry them, using them as ordinary cellular phones. For the very same reason, tourists have no problem in remembering to recharge the devices' batteries. And, while the information they gather may be of little worth in the context of micro-level research, it could prove useful for studies carried out on larger geographical scales (Figure 6.7D).

Experiment IV Comparison Between Different Types of GPS Receivers

The goal of this experiment was to compare the performances of four GPS receivers in the PortAventura amusement park (one hour south of Barcelona) in order to decide which type of device should be purchased for a tracking project that was slated to take place in the park (some of the project's findings are presented in chapter 8). This experiment reveals the meticulous process of choosing the most suitable equipment for a specific study, since even among devices of the same technology (GPS in this case) different qualities exist due to differences in hardware or software.

Four types of GPS receivers were tested (Figure 5.2.). Special consideration was given to the process of downloading the data obtained from the GPS receiver, as an important feature of the project was to make a real-time analysis of the participants' activity in the park (see next chapter). Below is the description of each of the devices data download characteristics:

Figure 6.8 Location kit.

1. Trackstick II. The data from this device can be downloaded only via USB cable connecting the device directly to a computer.
2. i-Blue 747 Bluetooth Data Logger GPS Receiver. Data from the device can be downloaded via USB cable or Bluetooth wireless protocol connecting the device directly to a computer.
3. GlobalSat BT-335. Data can be downloaded only via Bluetooth wireless protocol, which connects the device directly to a computer.
4. Royaltek RGM-3800 GPS Data Logger. The data can be downloaded only via USB cable connecting the device directly to a computer.

The method chosen for testing the different devices was to carry them along three walking routes that simulated visitors' time-space activity and the different indoor and outdoor environments in the park within which the GPS receivers would have to navigate.

Track 1

The goal of this track was to test that all of the receivers functioned correctly and obtained locations and then to assure that they were able to download the data in an efficient manner. This track included a quick walk into the amusement park and up to the park's internal "train" station, a wait at the "train" station, a ride on the "train" to the other end of the park, and a walk out of the park. The results obtained from this track include only three receivers; the BT-335 receiver did not function properly and did not save the route at all.

Track 2

This track included a complete walk in the amusement park. All four receivers functioned properly and gave results. While all of the receivers were programmed to save a point every 5 seconds, the BT-335 receiver saved a point every 30 seconds. On this track, one attraction ("Sea Odyssey") was visited. This indoor attraction is located in a closed, dark room without windows and is a very challenging environment for GPS.

Track 3

This track was traversed in the morning right as the park opened. The park was opened gradually and at two distinct points crowds waited to enter parts of the park that were not yet open. On this track, two attractions were visited: The "Grand Canyon Rapids," a ride in which one floats over a water course in a round boat while being splashed along the way, and a special Christmas enchanted forest (mostly indoors) in which one walks and meets different characters. All of the receivers functioned as planned and sampled locations every 5 seconds.

Conclusion of the Experiment

This test immediately eliminated two receivers based on the findings: The Trackstick II did not produce locations with the expected accuracy and therefore was not relevant for use in the planned project. Although it gave accurate locations, the RoyalTek receiver's power ran out after a few hours and it was therefore deemed unsuitable for the project as well.

The two remaining receivers performed well and gave high-quality locations; a number of misplaced points were obtained with both receivers and one unexplained incorrect path was obtained by the i-Blue. The i-Blue's design made it a better option for use in the project, having the on/off button placed on the side, so that it was less accessible than it would have been if placed on the front. The importance of this feature is that a participant can turn the device off if the on/off button is accessible; for research purposes it is better if the button cannot turn off easily.

Another aspect that made the i-Blue model more suitable for use in this project was its ability to download the obtained data using a USB cable and not Bluetooth communication; downloading data with a USB cable is simple and reliable. The BT-335 did not have the option of downloading data using a cable, relying solely on Bluetooth communication. That said, the BT-335 was slightly superior to the i-Blue and did not have any "misplaced" paths like the i-Blue did. In the end, the two models were purchased and used for the project; both of them functioned very well despite the different methods used to download the obtained data from the receivers to a computer.

Summary: Testing the Tracking Technologies

This first part of the chapter presented the results of four experiments, each of which used one or more of the three principal tracking technologies currently available. All of these technologies could potentially be used as effective tools for analyzing the spatial and temporal behavior of tourists, but only if, as previously noted, the tracking units used do not restrict or alter the subject's behavior in any way. Put simply, they must be fairly light, easy to carry, and able to track the subject reflexively without forcing him or her into taking any kind of special action.

In this respect, the land-based techniques have an advantage over the GPS, in that the end units do not need a direct line of sight to the sky, and therefore could obtain location in buildings. On the other hand, GPS devices have the advantage over land-based tracking methods when it comes to obtaining accurate data. This makes them a suitable means to be used in micro-level investigations, such as studies which record the number and density of tourists visiting historic cities, tourist attractions, theme parks, and similar locations, all of which demand high-resolution data.

There are limitations to the four experiments that were conducted and presented in this chapter. First, the locations for the tests were chosen due to their convenience, aside from the fact that they represented typical tourist destinations. Second, only one test was conducted in each location and with a particular device. This could lead to idiosyncrasies associated with the particular device or location.

Table 6.2 compares the GPS and land-based tracking technologies as a means of gathering data on tourists' time-space activity.

RESEARCH STRATEGIES FOR IMPLEMENTING TRACKING TECHNOLOGIES

The most important questions that influence the design of any study regarding the time-space activity of tourists in a destination are:

Table 6.2 A Comparison of the Principal Tracking Technologies

	GPS	TDOA	Cellular
Accuracy	High	Medium	Low
Availability	Worldwide	Only in areas with the appropriate infrastructure for TDOA.	Only in areas with cellular coverage (in areas of human activity and along transport networks).
Urban/ Rural	Works well in rural areas with an open terrain and where the sky is unobstructed.	High infrastructure costs means that it is economically viable only in urban environments, which contain a large number of potential users.	More accurate in urban areas where the density of the transceivers is higher.
Usage Costs	None. The satellite signals are free. As it is a one-way broadcasting system, adding more users to the system does not strain it in any way.	The cost of the service varies from place to place.	The cost of the service varies from place to place and according to the supplier and the billing plan.
Privacy	Privacy is not an issue; the ability to locate a user in the system is user-controlled.	User privacy is questionable; the system can locate a user at any time without the user's knowledge.	User privacy is questionable; the system can locate a user at any time without the user's knowledge.
Real Time	Traditional Global Positioning Systems cannot track in real time. In order to do so, a communication channel must be added to the system.	Allows tracking of subjects in real time.	Allows tracking of subjects in real time.
Radio Frequencies	Works at high frequencies, frequencies allocated specifically for GPS use throughout the world.	Uses lower frequencies, which can, in principle, also be used by other parties.	Uses lower frequencies, which can, in principle, also be used by other parties.

1. Scale: Questions of spatial and temporal scale, such as the geographic scale of the investigation, the type of destination under investigation, and the temporal scale of the study (i.e., how long the tourists have to participate in the research; different specifications for the tracking device that are necessary depending on whether it is a one-day or a week-long tracking process).

2. Population: The second set of questions relates to the nature of the tourists participating in the study: Who is being observed? Is it the whole tourist population or just a specific segment? Is it organized tourism or individual tourism? In the case of organized tourism, the tracking process is much easier, since the tour guide can be contacted well in advance of the group's arrival and, with his or her cooperation, information about the group can be obtained and tracking devices distributed in a much simpler manner than when working with individual tourists.

3. Nature of non-spatial data: What kind of information do the researchers expect to gather from a participant? Is it only his or her time-space activity or do they wish to gather information about his or her satisfaction during the visit to the destination? Researchers might be interested in the money expenditure of the visitor in time and space or in his or her feelings toward the tourist experiences he or she has been exposed to during the visit.

The answers to these three fundamental questions will have a direct impact on:

(A) Technology: Which tracking methods will be used in the study (land-based or satellite-based)? Should only tracking technologies be used, or should a combination of questionnaires (paper or digital, using, for example, Personal Digital Assistants) or interviews accompany the tracking? After choosing the main methods for data collection, various options exist regarding the specific type of tracking devices to be used—in terms of battery power, data-storing abilities, and other features that will be discussed later in this chapter.

(B) Point of sampling: Where should the tourists be sampled? Should it be at the entry point to the destination? At the hotel or another type of tourist accommodation? Maybe the sampling procedure should take place at an important attraction? In case of mass analysis of cellular phones, data collection will be done at the cellular operator offices. Each location for sampling tourists has it own advantages and weak points.

Thus, for each research design a thorough evaluation process should be undertaken in order to ensure that the best option in terms of tracking methodology and sampling location has been adopted.

Below we present various strategies for conducting tracking studies at different, yet typical, outdoor tourist destinations. They differ in scale, from the tourist attraction scale to the national or even continental one. It should be clear that this is not a closed list of options; we are sure that the imagination of other researchers will expand the possibilities, making the recommendations below richer and more complex in the future.

As this book does not replace other existing works on field methods in the social sciences, a number of issues will not be discussed in this chapter: We do not tackle issues of sampling, the topic of bias as a result of segments that do not wish to participate in a study, questions about the recruitment of participants, or the topic of incentives to participants. It should also be noted that various factors could have an impact on the level of participation in a study, such as cultural background and personality features; since this is also a general concern for any field study, we will not expand on the topic.

Enclosed Outdoor Environments

Theme parks and other large-scale paid attractions, such as zoos, open-air museums, natural attractions, archeological sites, and heritage sites (the Prague castle for example, is comprised of numerous buildings, and the tourist explores the entire site by walking from one to the next) are ideal locations for gathering a representative sample of visitors, since they generally have few entrances and exits. The fact that the visit is limited to the opening times of the attraction means that the maximum visit will usually not be longer than about twelve hours (in the case of the largest theme parks); the average visit will be much shorter, depending on the nature of the attraction and the entrance price—the more expensive the entrance fee is, the longer the stay of the visitors.

It is important to emphasize that in museums or other indoor environments, GPS receivers do not function and, due to the small geographic scale, the use of cellular phones for obtaining locations is irrelevant. As a result, the only viable tracking methods for such environments are RFID or Bluetooth technologies. In this case, transmitters can be placed in different locations in the building and the visitors can be given RFID tags or a Bluetooth device. When the visitor is detected by the RFID or Bluetooth transmitters, the time and duration of his or her presence in the location of that transmitter is recorded. Additional technologies that can be employed in such small-scale environments are closed-circuit television cameras (CCTV, as was mentioned in Chapter 3).

Place of Sampling

Paid attractions usually have limited and clearly defined entrances. The reason for the limited number of entrances has to do with operational cost saving; less staff operating entrances and exits and limiting the amenities that are part of such a complex—like a souvenir shop, for example—leads to greater efficiency and thus cost-saving. This fact is very convenient for implementing a tracking research project, since sampling the visitors entering is relatively easy. The visitors must stop at the entrance for the purchase of ticket and to gather information, therefore this is a convenient time to attempt to recruit them to a study. This is true for the end of the visit as well; a limited number of exits make it relatively easy to collect tracking devices when the guests finish their visits.

Suitable Tracking Technology

The relatively small size of these attractions, ranging from approximately ten acres to several hundred acres in most cases, makes GPS the most attractive option for tracking due to the high resolution that is essential in such a case. In smaller locations, using RFID tags could also be a viable option, though probably a more expensive one; the time-space data gathered would also be inferior to those obtained by GPS, since the data gathered will only indicate whether a visitor stayed in proximity to a RFID or Bluetooth transmitter (the visitor will have to carry a tag). Thus, in such a case, in order to map the mobility pattern of a visitor, a large number of RFID transmitters would have to be mounted; this would be significantly more expensive than giving each participant a GPS device to carry.

Another option, which is probably the optimal one, is to combine the advantages of the two technologies, giving visitors both a GPS receiver and an RFID tag. The GPS receiver can thus be used to obtain data about the whereabouts of the visitors in open spaces and the RFID tag can map patterns of activity inside small indoor environments. For example, when a visitor enters indoor attractions in a theme park or visits shops and restaurants, RFID transmitters placed in those small environments will increase the level of information gathered about their time-space activities.

The new generation of GPS loggers that has recently entered the market (2008) is of lower voltage, allowing the receivers to operate for longer without a change or charge of batteries. Their data storage capacity can hold approximately one week of data. There are, therefore, no technical barriers for this kind of implementation. Due to the relatively short period of tracking in this case, the tracked data can be retrieved from the tracking device at the end of the visit.

In the case of theme parks, which contain rides that may involve the visitor getting wet, it is recommended to supply a small waterproof pouch to the study participant in order to ensure that the device does not come into contact with water.

Challenges

A precondition for conducting research in such environments is the cooperation of the attraction's owners. In cases in which the owners have requested such research, this is, obviously, no problem; however, if the motivation is purely academic, the first obstacle in such a study will be to obtain permission and cooperation, which are not always granted.

Another concern in relatively small environments is the question of who is being tracked when distributing one tracking device to, for example, a family of two adults and two children. In family visits to theme parks there is a tendency to split for part of the visit due to different interests and capabilities (e.g., children of different ages). One possible solution is to distribute a GPS receiver to each member of the visiting group and then to analyze the different itineraries inside the park.

Description of Fieldwork in the PortAventura Theme Park

The PortAventura theme park is located in the Costa Dorada, near the resort town of Salou, about 100 kilometers south of Barcelona. Data collection was conducted in two rounds of one week each. The first stage took place during spring 2008, the second wave during the summer of the same year.

Visitors to the park were approached by employees from the park's market research department. After initial screening questions (the aim was to sample families with young children), the research staff asked potential participants if they wished to participate in the study and carry a GPS unit with them throughout their visit in order to assist the park in studying guest satisfaction in the context of their time-space activity in the park.

About 80 percent of the families that were suitable to participate in the study agreed to do so. The technical performance of the GPS devices was very good as well: Out of 288 families that took part in the research, 277 families were included in the final analysis (96%).

Two different types of data were collected for each family using two different data collection methods: (1) The visitors' socio-demographic and personal data were collected at the park entrance using the park's regular visitor profile questionnaire. (2) Time-space data were collected by GPS devices that were set to sample the location of the visitors every ten seconds.

When participants finished their visit to the park and returned their GPS devices to the designated place near the exit they were asked to participate in an interview regarding their spatial activities in the park. Thanks to a real-time analysis of their GPS track (see Chapter 7) the interviewers could ask them specific questions about their day in the park. The average interview took about 15 minutes. In total, 254 interviews took place, reflecting a very high level of compliance on the part of the participants.

One of the purposes of this pilot project was to gauge whether visitors to the park would be willing to participate in GPS tracking research and, more importantly, to agree to be interviewed at the end of a long day at the park. The results were very positive, as described above. Furthermore, these results were achieved without any incentive being promised to participants. At the end of the interviews a modest gift was given. It is possible that the participants expected to receive something at the end of the study; however, this was not discussed or promised at any point.

Historic Cities

"Historic cities" refers to historic sites that are primarily visited by daytrippers and not by tourists. The size of these locations may be very similar to the previous type of tourist destinations; however, the main difference between the two categories for researchers hoping to engage in a study of time-space patterns is the fact that historic cities do not have a defined entrance point; access is possible from various directions.

Place of Sampling

The result of the lack of a defined entry and exit point is a complication for the sampling process and leads to a search for more possible locations for sampling the visitors. As a consequence, in order to receive a representative sample, more than one sampling point might be necessary. Places for sampling could include the city's tourist information center, especially in cases in which it is located at an entry point to the city—for example, in a train station. In each destination, different circumstances could influence the places of sampling.

Suitable Tracking Technology

Due to the small size of this type of destination, GPS technology would be the most suitable for use in a study. However the narrow alleys that are typical to historic cities could create a challenge for obtaining GPS signals, though the intense data collection by the devices (ranging from once per second to lower resolutions) means that the GPS signal will be found again by the device once a line of sight to the satellites is restored. Some of the experiments presented in the first part of the chapter took place in historic cities and the results demonstrated good reception of the GPS receivers in those dense environments, generally speaking.

The tracking of daytrippers in historic cities does not create a challenge in terms of data storage or battery life of modern GPS devices. It is assumed that a day visitor will carry the device for an entire day and then return it to the location of sampling he or she visited at the beginning of the day. In cases in which tourists are sampled at their hotels for more than one day, there are various complications; they will be dealt with later on in this chapter.

Challenges

The main challenge in this type of destination is where to sample the visitors in the absence of a defined entry or exit point. This concern exists naturally when larger cities are been investigated. When resources are available, sampling at the different entry points could tackle this problem. Alternatively, sampling can be done at a main entry point. In the case described below, we chose to sample the tourists at the visitors' center. Naturally, the conclusions of the study are limited to visitors who began their trip in the visitors' center, but this is an example of a pragmatic solution to this complex issue.

Description of Fieldwork in the Old City of Akko

The location kit (Figure 6.8), which was distributed to each of the individuals taking part in the study, consisted of a GPS receiver and a Pocket PC. The GPS receiver transmitted the visitor's position to the Pocket PC via Bluetooth technology; the PC then logged the location's coordinates. This complicated constellation was created, since GPS loggers were hard—or impossible—to find in 2004 when this project took place.

Aside from carrying GPS receivers, the participants were asked to complete questionnaires before and after their visits. The questionnaires collected information regarding the visitors' socio-economic characteristics, the nature of their specific tour (place of stay in northern Israel, time allocated for the visit to Akko, whether it was their first visit to the destination, the use of a city guide or map, etc.), and finally their perceptions about the nature of Old Akko as a locality and a tourist site.

In order to obtain a sample group of visitors, a stand was erected alongside the Old City's visitors' center's information booth. The center itself is located at the entrance to the underground Crusader City, which, as the city's most celebrated attraction, is usually the first site most visitors elect to visit. Only visitors who had purchased tickets to the archaeological sites run by the Old Acre Development Company Ltd. were asked to participate in the study. Of these, the few who had already spent some time in the Old City were immediately rejected as viable subjects, since their tracks would not represent their entire visit. The disadvantage of the location selected for sampling was that visitors to the Old City who did not come to the visitors' center were not represented in the study.

The data were collected between the months of June and August 2004 on 19 non-consecutive days. During this time, a total of 246 visitors' tracks were obtained, of which 107 tracks were discarded, for the following reasons: (a) technical reasons, for example, cases in which the location kit failed to log the full track of the visit (n=38); (b) as a result of subjects failing to complete the questionnaire (n=23); (c) owing to subjects deciding to return the location device before completing their tour of the Old City (n=46). In comparison to the fieldwork described in PortAventura (in 2008), the number of participants whose information was not used for the study was relatively high. This is explained by the difference in tracking devices, which improved significantly in the four years between the studies, as well as a higher quality of research design.

NATURAL PARKS AND WILDLIFE RESERVATIONS

Destinations of this sort are basically expansions of the enclosed outdoor environments. Some ski resorts could be included in this category as well, in cases in which they have well-defined entry points. For example, Wengen, a car-free ski resort located not far from Interlaken in Switzerland, would be included in this category. Tourists who wish to reach the resort must park their cars in a large parking lot in Lauterbrunnen and take a special mountain train to the resort.

Place of Sampling

Locations included in this category usually have limited, defined, and clearly defined points of entry, such as visitors' centers, where tourists can be sampled. In cases in which there is more than one such point, sampling can be done in such a way that this feature is taken into account.

Suitable Tracking Technology

These locations are ideal for GPS-based research. For one thing, they are mainly outdoors. Cellular tracking, as we saw, cannot supply accurate locations in rural areas because, unlike in cities, there is a relatively low density of transceivers and, therefore, the spatial resolution is very low: about five to ten kilometers (see Mateos and Fisher 2006). GPS technology is thus the best option, but there is a need to transfer the data from the GPS device via an external communication protocol (for example, via SMS or GPRS). Alternatively, the device could have the ability to store the obtained locations in time and space on an internal hard disk or memory card for later data retrieval. However, this option is problematic; in case of malfunction, all data obtained are lost.

Challenges

As the tracking period in these kind of destinations becomes in many cases longer than one day, the researcher must face questions regarding the compliance of the participants: Did they remember to charge the tracking device? Did they remember to take the device with them? Might they have left it in their hotel room or car? These concerns are relevant also with the traditional methods discussed in Chapter 3.

One method for tackling this potential problem is to ask participants to keep a diary. The researcher thus gets an idea about their time-space patterns and can compare it to the data obtained by the tracking device. Another option is described, in the next section of this chapter.

When several entry and exit points exist, the participants must return the device at a sampling point and—equally importantly—be interviewed or fill in a questionnaire.

Large Multifunctional Cities and Large-scale Regions

This is the most complex category in terms of data collection regarding time-space activities of tourists. This holds true not only for the more groundbreaking research methods but also for the different traditional methods described earlier in this book. The reason for the difficulty is the fact that, due to their size and complexity, larger-scale regions and cities do not have one main entry point and are frequented by several types of visitors. One research strategy could be to investigate just one kind of segment of tourism in a city or a region, for example, cruise ship passengers who disembark in a city's cruise terminal. In this case, sampling would be relatively easy, but the data collected would necessarily be limited, referring to just one of the many segments of tourists in the city. Other options will be described below.

Place of Sampling

Three possible locations exist for sampling tourists in a large multifunctional city or tourist region: tourist sites, hotels (or other types of guest accommodations), and entry points to a city or region. Each of these has its own advantages and disadvantages.

Tourist sites: Sampling at tourist sites has a bias that is created due to the fact that tourists who do not visit those sites are not represented in the sample.

Hotels and guest accommodations: The principal advantage of this type of sampling is that the tourist receives a device upon arrival and can return it at the hotel upon leaving. However, there are a number of challenges to this method. Firstly, in order to achieve a sample that represents the overall tourism population, a variety of hotels should be included as sampling points. The second challenge is that not all tourists choose to stay in commercial accommodations; some may choose other accommodations (private homes or religious or educational institutions, for example) and others may tour a city for only one day. A final problem is that a tourist's checking out of a hotel does not necessarily indicate that his or her trip is over; he or she may still return to the region or city, or relocate to a different hotel during the visit for a variety of reasons.

The hotel's cooperation is also a question in research of this sort. It cannot be taken for granted, though it should be noted that a two-year study currently taking place in Hong Kong (2008–2010), in which tracking devices are distributed to tourists in a number of hotels of varying levels and in various locations, proves that it can be successful.

Points of entry and exit for the city/region: This method has a number of significant advantages: 1) All tourists are represented, even those not using established accommodation services. As such, a truly representational sample of all tourists in the destination is created and 2) The fieldwork is efficient in that, using only a few points, all tourists can be sampled. In addition, it is possible to concentrate a number of researchers who speak different languages in one location at one time, thereby broadening the population represented.

The first possible sampling point, which comes to mind intuitively, is the airport. A number of challenges arise: Consent on the part of the management must be attained, which cannot be taken for granted. In addition, tourists reaching a destination generally wish to leave the airport as quickly as possible; often, they are tired from their flight and thus one can assume that it will be difficult to enlist them for the purposes of a study.

Attractive locations at the airport for enlisting potential subjects include information booths and car rental stands. These are places that tourists reach once they have collected their luggage and are setting off toward their destination. However, this does limit the sample of subjects for the study. On the other hand, if the study focuses on tourists traversing the destination in rented cars, this is an ideal methodology for the distribution of tracking devices and the completion of questionnaires.

Aside from airports, border crossings can serve as possible points of enlistment for studies (provided such crossings exist); even if border crossings are not defined, visitors' centers located near borders can be suitable as well.

Yet another possibility to overcome the challenges mentioned can be the enlistment of subjects for a study at their point of origin before they have left for vacation. In this case, the picture of the tourists' spatial activity during their vacation is the fullest. Researchers can, for instance, speak to potential subjects by their gate at the airport, assuming that they have already undergone security checks and have free time before getting on their flight. To the best of our knowledge, this method has never been used; proper incentive would need to be given in order to enlist volunteers for a study of this kind. However, there are many advantages to this type of study.

Suitable Tracking Technology

When tracking a visitor to a large region or large city, cellular phone tracking should be taken into consideration as an option despite its relatively low spatial resolution; in some cases, the research design might enable more flexibility with this parameter. It is also possible to use cellular phones with embedded GPS receivers in order to gain higher accuracies. However, as was demonstrated in the first part of this chapter, those GPS devices tend not to function as well as GPS devices designed with data collection and storage in mind.

The advantage of a cellular phone is that it can transfer data to the researcher over the cellular network from time to time. This is important since, if the device malfunctions, the researcher knows in advance and can even try to contact the participant in order to solve the problem.

It is important to explain here that we do not suggest using the tourist's private cellular phone for tracking. He or she would be given a cellular phone for the research period itself. This is the best option in terms of visitor privacy and enabling the tracking. We recommend charging the phone with "air time," providing the tourist with incentive to participate in the study and carry the device throughout his or her visit. Another related advantage is the probability that the participant will remember to charge the cellular phone on a regular basis, given that it is a device he or she uses for purposes other than the tracking.

GPS devices can be relevant on this scale as well, but the question of the duration of the tracking period arises. If the intent is to track one day of a visit in a city then simple devices will do—as described previously in this chapter—but if the tracking is intended for the whole period of stay in a city or region, which could be from several days to several weeks, the devices should have significant data storage abilities, or, even better, a way to transfer the data to the researchers during the tracking itself. For example, a GSM modem using SMS or GPRS data transfer protocol can be used. Several devices on the market already offer such features.

Challenges

The main challenge in relatively long participation periods is the memory of the participant. Even when motivated to participate, after several days there is a high probability that he or she will forget to charge the tracking device or simply forget to take it during the day. Just as with time-space diaries, the researchers must rely on the participant to carry the device at all times. When reviewing GPS data, the researcher must ask: What data am I missing? Is there a common denominator in the data that I am missing? Is there a certain profile of subject that is not willing to participate?

When spatial data are collected using traditional GPS receivers, these questions remain unanswered; the researcher has no way of knowing whether the receiver was left behind and, if so, why. The only way to answer these questions is to ask the participants, who may not want or be able to give accurate answers.

Understanding the nature and distribution of missing data is extremely important in order to avoid information bias. For example, if the GPS were located in the participant's hotel from six o'clock every evening until the next morning, this could mean that the participant does not leave his or her hotel in the evening, exhausted from touring the whole day; it could also mean that the participant spends his or her nights partying in clubs but prefers not to take the tracking device. In that case the data on the whereabouts of the participant are not only missing, they are actually incorrect, resulting in systematic error in the data set. In sum, while compared to more traditional time-space diaries GPS devices have the potential for more complete data collection, their use is also subject to the same potential biases.

At this point, we wish to present a method that was implemented in another type of research—tracking elderly people—but could potentially also be used in tourism studies and could help overcome this challenge. "Sen-Tra," a five-year Israeli-German cooperative research project, examines the outdoor activities of elderly cognitively impaired persons by taking advantage of tracking technologies (for more details, see Shoval et al. 2008).

The GPS tracking kit (obtained from HomeFree Wireless TelehomeCare Solutions, Tel Aviv, Israel) consists of three main elements: 1) a unit that is carried and contains a GPS receiver, GSM modem, and RF receiver; 2) a device resembling a wristwatch, which contains an RF transmitter and additional sensors, including one that detects whether the watch is being worn on a person's body; and 3) a stationary home unit that repeats the RF signal and allows the participant to walk freely around the house. The waterproof "wristwatch" RF transmitter tells the researcher whether participants are complying with the study's guidelines and carrying the GPS device at a given moment. If the strap is opened or the RF transmitter is not in contact with the body, the system issues a notification. If the watch is worn on the hand, but is further than ten meters from the GPS receiver (at home this maximum distance is seventy meters because of the home monitoring unit), a notification is issued again. These notifications allow the researchers to determine whether the research subject is participating at a satisfactory level. Aggregating these data enables the researcher to measure the participant's overall level of compliance and, as a result, to assess the quality of the data obtained during the relatively long tracking period.

CONCLUSION

Is technology that monitors the level of the participation by using what amounts to an electronic bracelet feasible for tourists? Or does it place too great a demand on people who wish to enjoy their vacation? This is an important question that has yet to be answered. At the same time, it is also clear that the whole range of possible incentives can be used in any type of tourism study in order to form the best research design, find the ideal place for sampling tourists, and employ the most suitable technology for a study.

Do the tracking devices return at the end of the tracking period? This is one of the frequently asked questions researchers face regarding the implementation of this method for research. The answer is that this must be taken into account when creating a research design involving this technology. Participants' contact information should be taken in case they forget to return the device and must be contacted. However two points should be made on the subject: The first is that, with a cost of about $100 for the best available GPS logger on the market today, even if some devices are not returned, this is not fatal for a research project. The second point is that, in our experience in various tracking projects in the field of tourism, we have yet to lose one tracking device. A Danish team that investigated the activity of more than one thousand visitors in four parks reported that they lost only three devices during the overall period of tracking in the field (Hovgesen et al. 2008).

Table 6.3 below summarizes the different points made throughout the chapter regarding tracking tourists in different types of tourist destinations.

The table clearly indicates that (at the time of writing) the most suitable technology for use in tourism tracking projects is the GPS logger due to its low cost, high reliability, ability to store data, and—perhaps most importantly—flexibility in application to different scales of tourism research.

The next two chapters continue this line of thought, presenting the information and knowledge about the tourist and the destination that researchers can expect to obtain by using the technologies described in this chapter.

Table 6.3 Comparison of Different Research Environments

Challenges	Recommended Technology	Place of Sampling	
(1) Researchers must obtain the consent and cooperation of the company running the attraction. (2) The researcher must define which of the visitors is being tracked (e.g., in the case of a family with children).	Mainly GPS loggers. In small attractions, the use of RFID or Bluetooth could be possible as well, but the combination of the two methods is ideal.	Near the entrance/s to the destination.	Enclosed Outdoor Environments
When a defined point of entry does not exist.	GPS loggers.	Entry points to historic city / Old City in larger cities, such as tourist information centers, train stations, etc.	Historic Cities
Questions of tourists' compliance in relatively long tracking periods.	GPS loggers, but with the option to transfer the data via a communication protocol such as SMS or GPRS.	Near the main entrance/s to the destination.	Natural Parks and Wildlife Reservations
Questions of tourists' compliance in relatively long tracking periods and finding a suitable place for sampling that will reflect the tourist population.	Cellular phones with or without embedded GPS function; GPS devices with the ability to transfer obtained data via a communication protocol such as SMS or GPRS.	Airports, border crossings, hotels, important attractions.	Large Multifunctional Cities and Large-scale Regions

7   Understanding the Tourist

The mapping and modeling of tourists' spatial activity is viewed by many researchers as an under-researched field in which much progress is still needed (Fennel 1996, Pearce 1987, and Prideaux 2000; Modsching et al. 2008). Advances in the development of tracking technologies offer an opportunity to further and deepen research of this nature. Where has the tourist been? How long did he or she stay at each site? What mode of transportation was used in order to get to the site? All of these questions can be addressed using data collected by tracking technologies, which detail the spatial behavior of visitors to a destination. Furthering the understanding of these subjects allows planers and policy makers to make informed decisions regarding policy and to address tourism development in a more informed manner. Deepening the understanding of tourists' spatial activity offers the researcher the ability to:

- Analyze the influence of the duration of visits (time budgets) on the spatial behavior of visitors;
- Ask the visitor questions regarding his or her spatial activity at a very high resolution;
- Examine the use of means of transportation (pedestrian, private transportation, public transportation);
- Identify general patterns of spatial behavior of visitors and categorize them according to socio-demographic characteristics; and
- Create typologies of tourists based on their spatial behavior and enrich non-spatial typologies by characterizing types of tourists' spatial activity.

The goal of this chapter is to introduce the possibilities that tracking technologies offer for deepening our understanding of the spatial behavior of tourists within a destination. The data presented were collected using GPS. This is chiefly due to the fact that (at the time of writing) it was found to be the most suitable for research, as explained in the previous chapter.

This chapter begins by describing what can be learned from analyzing data collected from one tourist on one track. It discusses possibilities for real-time analysis and the integration of spatial data into tourist questionnaires and interviews. An explanation of how spatial data can be manipulated to create different variables that can be incorporated in traditional research methods follows.

The second section of the chapter discusses typologies of tourists. These typologies are, in many cases, created using spatial data by applying a sequence alignment method that originates in biochemistry; in other cases, typologies are created without using spatial data but are enriched by adding the understanding of the spatial activity of these types of tourists.

WHAT CAN BE LEARNED FROM A GPS TRACK: A TALE OF ONE TOURIST IN HEIDELBERG

Heidelberg is one of Germany's most popular tourist destinations, its historic center and castle drawing tourists from all over Germany and the world. In 2004, over three million people visited the town, its hotels registering over nine hundred thousand person-night stays (Freytag 2002). On May 23, 2005, one of the book's authors traveled to Heidelberg. During his visit, the author, acting as the experiment's subject (and henceforth referred to as such), carried with him, in a specially designed harness, an Emtac CruxII BlueTooth GPS receiver and Pocket PC (see image of the kit in Chapter 6, Figure 6.8). The receiver, set to record one tracking point per second, was secured by a strap situated just below the shoulder, saving the subject not only from having to hold it in his hands, but also from having to constantly manipulate it in order to ensure that it was exposed to the sky. Using wireless Bluetooth technology, the data from this essentially passive receiver was transferred instantly to the Pocket PC.

It should be noted that GPS technology has developed so rapidly over the last few years that today the storage of obtained data is possible within very small GPS data loggers that are efficient in terms of data storage and energy consumption and therefore the use of such a large kit consisting of several elements (the GPS receiver and a PDA) to store the collected data is no longer necessary; thus, in current studies, the burden for the participant is significantly decreased.

The tracked tour of Heidelberg lasted four hours, during which the subject covered 19.3 kilometers. The subject began his journey, accompanied by his host, at the Ibis hotel, situated alongside the town's main train station (see respective numbers below in Figure 7.1).

1. Leaving the hotel, the two drove to the Heiligenberg Mountain, where they disembarked and climbed an observation tower for a panoramic view of the Old Town's skyline.
2. Returning to their car, they took a brief drive up the mountain, parked, and climbed a further 65 meters to the Heiligenberg's summit.
3. They then traveled back to town, driving to the Im Neuenheim Feld district and the University of Heidelberg's new campus. They parked the car on campus and took a bus to the Old Town.

4. Getting off at Bismarck Platz, the subject and his host entered Heidelberg's Old Town through its main gateway.
5. From this point onwards, they continued on foot. Having strolled along the Haupstrasse, the Old Town's main commercial street, the two made their way along several of its smaller side streets.
6. After 28 minutes of wandering, they stopped at the Café Burkhardt for 1 hour and 27 minutes.
7. At the end of their walk they boarded a bus, arriving at
8 The Ibis hotel and the trip's starting point (1).

The fact that in the course of his expedition in and around Heidelberg the subject used three different modes of transport—a car, a bus, and his own

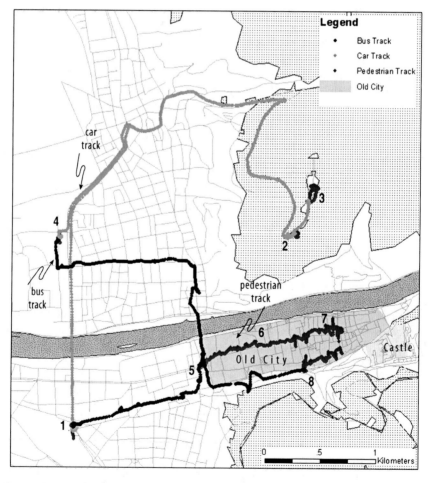

Figure 7.1 Track of tourist visiting Heidelberg.

two feet—is clearly marked on the track obtained using the GPS device (Figure 7.1). As noted earlier, until recently researchers have tended to focus on tracking vehicles, it being much easier to mount a tracking device on a car. Furthermore, unlike pedestrians, cars have a constant built-in power supply. Thus, the pedestrian alternative has, until now, been dependent upon the subjects carrying not only unwieldy devices but also a hefty supply of batteries, and needing to remember to change the batteries. Yet the fact that the subject in this experiment, like most tourists, switched his means of transportation in mid-journey underlines the importance of tracking the tourist and not the vehicle he or she might use. The development of smaller, more efficient batteries for use in portable electrical devices, in addition to the fact that these devices have over the last few years become smaller and lighter, means that it is now more feasible to track the pedestrians themselves.

The Old City of Heidelberg, with its many narrow streets and alleyways, is ideally suited for testing the GPS's ability to provide a clear and accurate track in dense urban environments. As the results of the experiment show, the GPS receiver remained fully functional throughout the test and was able to pin down the subject's position with remarkable accuracy. Indeed, neither Heidelberg's dense, maze-like streets nor the roofs of the bus or car used by the subject were sufficient to render it inoperative.

One point worth mentioning is that the density of the tracking points obtained varied according to the type of transportation used. This is not unexpected, as it is the speed at which the subject moves that determines the density of the tracking points. Accordingly, if, as was the case here, the sampling rate (e.g., one point per second) remains the same, the more closely packed the points are, the slower the motion and vice versa. This enables a researcher to establish whether the subject was traveling by car or moving on foot.

REAL-TIME ANALYSIS

A single track like the one described above can be analyzed in real time and supply valuable insight into a tourists' spatial activity for use in end-of-visit interviews. This data can be integrated into the interview process as a tool, helping the participant recall what he or she did throughout the visit, leading to a more meaningful and in-depth interview. Showing the person where he or she was and how much time was spent in each location can help walk the visitor though the events of the visit and enable the interviewer to ask questions that have a very high spatial resolution. An interviewer might ask: Why did you stop at this corner? Why did you walk out of this restaurant without sitting down to eat lunch? What made you choose this direction when you reached a certain junction?

Different methods can be employed to utilize the collected data. The most basic method is to present the collected information on a map. This can be done using Google Earth or another GIS. Many GPS loggers support automatic mapping of the collected data on Google Earth. In such a case, once the logger is connected to the computer the interviewer chooses the option of displaying the data and the data are imported seamlessly without any need for geographic knowledge. Many GISs, such as Google Earth, allow for the displayed data to be "played" according to the time of the samples. The participant can be shown the movie while being interviewed, stopping for the interviewer to ask questions regarding a specific time and location.

Another option is an automated analysis process run on the data; the interviewer can then use the results received to create questions for the participant. The automatic analysis can include calculating how much time was spent in different types of locations or the sequence of the visit in different locations. An interviewer might ask the participant about the reasons behind choices that were made throughout the visit or about attractions that were not visited.

Below is such a Table (7.1), produced when a visitor returned at the end of a day spent in the PortAventura amusement park. The table includes the different locations visited by the subject. The locations are categorized by type of location—attractions, entertainment, and shops. For each location both the number of visits, the total duration of the visits, and the average duration of visits to that location are presented in an hour:minute:second format.

Automatic analysis can also make possible the creation of a customized questionnaire. The questionnaire can be compiled by combining a set of rules that govern the inclusion or exclusion of questions. Designing a customized list of questions can minimize the length of a questionnaire that

Table 7.1 Results of Automated Data Analysis

Compound:	Number of Visits	Duration	Avg. per visit
A–Diablo	1	00:13:30	00:13:30
A–Furius Baco	1	01:05:12	01:05:12
A–Templo del Fuego	1	00:10:13	00:10:13
E–Long Branch Saloon	1	00:01:00	00:01:00
E–Temple Magic Jing-Chou	1	00:24:03	00:24:03
E–The Western Stunt Show	1	00:36:10	00:36:10
T–General Store	3	00:08:30	00:02:50
T–Records PortAventura	1	00:00:30	00:00:30
T–Western Clothing Company	1	00:03:20	00:03:20

participants are asked to fill out, leading to higher compliance rates and to more effective data collection.

This approach to research addresses the challenge of carrying out a combined analysis of tracking data and face-to-face interviews.

CREATING STATISTICAL VARIABLES
USING TRACKING DATA

The tracks collected from tourists using tracking technologies can be used to calculate many different variables that describe the spatial activity of visitors. Variables describing movement include speed, mode of transport, number of tracks taken, and the length of the tracks. Variables describing the locations that were visited by the tourist include the number of visits to the location, the timing of each visit, and the duration of each visit. Table 7.2 demonstrates the type of information a researcher can collect based on the spatial data he or she receives.

The creation of numeric variables from spatial data makes it possible to integrate spatial information into statistical analysis. Taking spatial data that are very descriptive in nature and turning them into quantitative data that can be used as part of statistical analysis creates a bridge between the spatial data and the non-spatial data, allowing the researcher to start developing theories that are broader and more inclusive than the theories that were developed without using spatial data.

An example of this kind of richness can be an addition to traditional visitor satisfaction surveys. Once data describing the visitors' spatial behavior are collected, it can be integrated with visitor satisfaction surveys; then, the researcher can ask a number of questions: In what way are the satisfied visitors' spatial patterns similar? Is there a spatial pattern of a satisfied visitor? Is it possible that a particular attraction was responsible for making the visitors agitated or unhappy? Perhaps families that ate lunch before visiting an attraction were calmer and therefore able to enjoy their experience to a greater degree? Answers to these questions can all be found in the time-space data collected.

Variables that are derived from the time-space data can also be of great assistance to planners, marketers, and policy makers. How many sites were the tourists able to visit in a day? What modes of transportation were used

Table 7.2 Example of Variables Derived From Tracking Data

Participant ID	Avg. Walking Time (hours)	Avg. Walking Speed (km/h)	Number of Attractions Visited
4567	2.45	4.6	12
1223	1.2	3.4	9

and at what times of day? Did visitors go back their hotels during the day? If they did, for how long did they rest? Who were the visitors who did not return to their rooms during the day? How much time did visitors spend out of their hotel? Do these figures differ according to the season?

Figure 7.2 shows the average distance of a visitor from the hotel in which he or she is staying at each hour of the day. The circle represents the twenty-four-hour cycle of a day, and the distance from the center of the circle represents the distance from the hotel. Thus, the figure shows that the visitor left his hotel between nine and eleven in the morning but did not venture out to a distant location. Perhaps he ate breakfast at a café near the hotel. Later in the day, at one p.m., the visitor spent an average of five hours away from his hotel, possibly visiting attractions or walking around the city, returning to the hotel at around six p.m. Later, at eight p.m., the visitor left his hotel again and traveled about 1.5 kilometers from the hotel, returning shortly before eleven p.m.

TOURIST TYPOLOGIES

The division of tourists into typological groups is, like any other typology in the social sciences or, indeed, any other areas of research, an essential first step in scientific investigation. This is no less true in the investigation

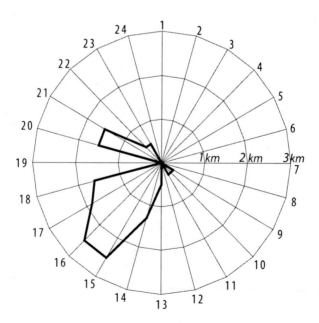

Figure 7.2 Average distance of a visitor from the hotel every hour of the day.

of the spatial activity of tourists. The remainder of this chapter will discuss the contribution of high-resolution data obtained by tracking technologies to the construction of tourist typologies.

There are two main approaches to typologies of tourists: spatial typologies and non-spatial typologies.

Spatial typologies divide tourists into groups based on the way in which they consume the space of the visited location. These typologies can then use non-spatial data such as demographic characteristics, personality traits, and data that describe the trip to seek a common trait that might be the reason for the different visitors' similar behavior.

Non-spatial typologies use spatial data to add depth and richness to their divisions by asking whether the different types of tourists are connected to each other not only by their characteristics but by their spatial activity as well.

Non-spatial Typologies

To date, two main approaches have been proposed in tourism research for the creation of typologies of tourists based on non-spatial characteristics:

1. Interaction typologies, which emphasize the nature of the interaction between the tourist and the tourist's destination. The best known of these typologies was proposed by Cohen (1972), who classified tourists into four main groups based on two criteria: one, the degree to which they seek out familiar as opposed to unfamiliar travel destinations, and the other, the degree to which they subscribe to mass tourism as opposed to individual tourism (for a more detailed classification based on the work of Cohen, see Smith 1977).

2. Typologies based on the analysis of the personality structure of the tourist. According to Plog (1973; 1987), American tourists are distributed "normally" along a continuum based on personality type: At one extreme, the "psychocentrics," who tend to worry and are characterized as self-centered, non-adventurous, and concerned with the more trivial problems they encounter; at the other extreme, the "allocentrics" who are characterized by self-confidence, curiosity, and adventurousness. These people see traveling to different destinations as a way to satisfy their natural curiosity. Plog goes on to claim that the different personality structures influence the choice of destination of holidaymakers. Additional typologies on the basis of the differences in personality structure have also been developed (see, for instance, Cohen 1979; American Express 1989).

Despite the centrality of these two theoretical approaches in tourism research, they explain only a small part of the tourist industry, namely, tourism for cultural consumption, tours, and visits. They ignore other significant segments of tourism, which apply mainly to the urban space, such as business tourism and visiting friends and relatives. This gap stems from the fact that these approaches adopt the sociological definition of tourism, which does not include these segments as part of its concept of "tourism." Similarly, the typologies that were created in the universe of sociology and psychology focus naturally on the tourist (the person) and, at most, on his or her connection with the host society. In other words, they see the space in which the tourist acts as uniform from a geographical point of view (see Figure 7.3).

In sum, there is a clear lack of theoretical frameworks dealing with the spatial activity of tourists at the tourist destination. The few studies that have dealt with the spatial activity of tourists in are descriptive, and, more often than not, even studies with an empirical base (such as Chadefaud 1981; Lew 1986; Shaw et al. 1990; Murphy 1992; Dietvorst 1994; Montanari and Muscarà 1995) employed only small samples and therefore

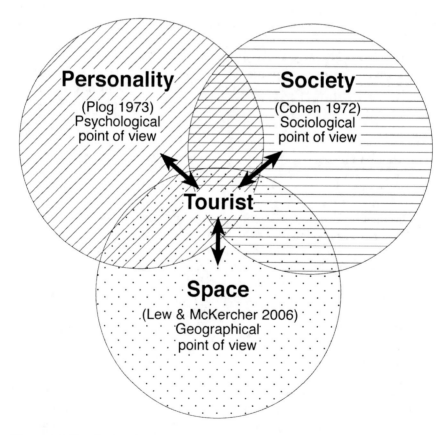

Figure 7.3 Typologies of tourists.

cannot deal systematically with the combined factors that influence the spatial activity of tourists in destinations.

Combining Data on Spatial Activity with Non-spatial Typologies

Figure 7.4 shows the distribution of a visitors' time budget as it was spent in different types of attractions throughout an amusement park. This analysis was done using data collected by GPS receivers that were carried by visitors to the park. Figure A illustrates the way visitors who had young children with them spent their time and Figure B presents how visitors without young children spent their time. Together, they demonstrate how researchers can see whether the spatial activity of two types of visitors who differ from one another based on one non-spatial criterion are also spatially different. An examination of the figure reveals that families with young children spend less time on rides but more time in shows and restaurants. This information can be very useful for the park's management, which can schedule more frequent shows on days when many families with young children visit the park (such as school holidays) and make sure to keep rides operational on days when the visitors are mainly visiting without young children.

Spatial Typologies

Pioneering work on spatial typologies of tourists was done by Lew and McKercher (2006), who published a paper establishing the theoretical approach to the modeling of patterns of spatial activity of visitors within a destination. In their paper, they presented two types of models that can explain the spatial activity of visitors in destinations, connecting the typologies to different itinerary types. The models that they presented were later used by Connell and Page (2007) as the theoretical background for a study of spatial patterns of car-based tourists in Loch Lomond and Trossachs National Park in Scotland.

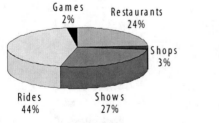
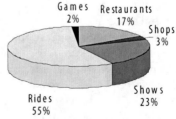

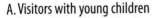
A. Visitors with young children   B. Visitors without young children

Figure 7.4 Distribution of a visitor's time budget in different types of attractions throughout an amusement park.

The first model (see Figure 7.5A) is a territory-based one; it refers to the variation in the distance that a visitor ventures from his or her place of accommodation. The distance varies from zero, when the visitor stays confined to the hotel, to unrestricted movement throughout the destination. Lew and McKercher state that the variation between visitors' behavior when classified according to this model can be related to the tourists' characteristics, time budget, motivations, interests, composition, and knowledge of the destination.

The second model (see Figure 7.5B) that Lew and McKercher suggest is a linear model that aims at capturing the different types of paths that visitors take within the destination or the different types of itineraries that lead to distinct spatial patterns. This model differentiates between point-to-point movement, loop movement, and movement that is a combination of both loops and point-to-point movement. This model is strongly influenced by the layout of the attractions at the destination but is independent of the mode of transportation, meaning that all of the spatial patterns can be achieved using different modes of transportation.

These typologies were developed deductively using theoretical materials and studies; they are not as strongly based on empirical data. The validation of these models and adding to their diversity and richness has been done in a limited manner but has yet to be done extensively. Data collected using tracking technologies will allow for the validation of these models as well as tailoring them to match the conditions presented at destinations of different nature.

Visualization Methods

When attempting to develop a typology it is essential to identify principles that can be used to divide the samples into groups that are distinct from one another in at least one way. Developing a spatial typology entails discerning a way in which the samples can be divided based on their spatial patterns. For this to be achieved it is necessary to present the spatial data in a way that will enable the identification of spatial patterns.

One of the ways to divide samples of time-space activity into groups based on spatial patterns is by mapping or visualizing the samples and then dividing the samples based on the resulting map or plot. The example shown here uses time-space data collected in Akko. These methods mainly use three-dimensional plotting abilities made available by modern geographic information systems; in this case, ArcScene, developed by ESRI, was used.

Nestled along the Mediterranean shore, Akko is one of the oldest continuously inhabited towns in the world. As we saw earlier, it is most famously known for its extraordinarily well-preserved underground Crusader City and its Old City, which boasts numerous other interesting archeological sites, plus a working fishermen's wharf and a typically Middle Eastern market (Figure 7.6).

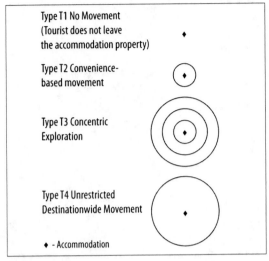

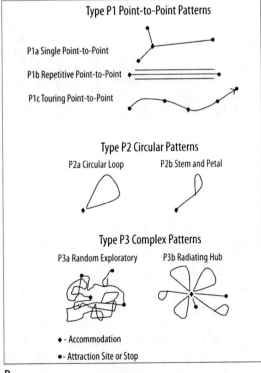

Figure 7.5 Models of spatial activity based on Lew and McKercher 2006.

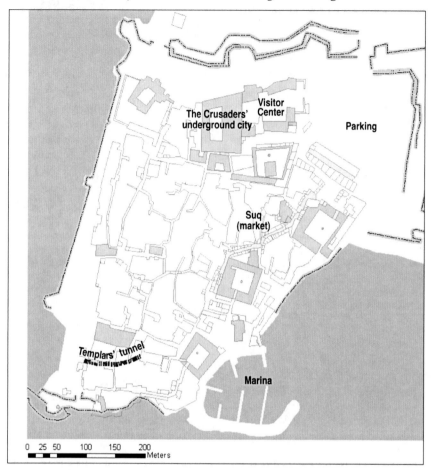

Figure 7.6 Map of Old Akko.

Visual methods that assist in the analysis of time-space data can be divided into two main categories: two-dimensional geovisualization and three-dimensional geovisualization. The two-dimensional methods plot the time-space data on maps using both spatial and temporal scales; for example, behavior in time and space can be represented by plotting one dot every minute. Three-dimensional methods that take after Hägerstrand's time-space prisms use the additional dimension to represent time.

Using three-dimensional Geographic Information Systems to create time-space prisms as discussed in the Chapter 3 opens new options for analysis and aggregation of time-space data. These new options have been made possible by the introduction of computerized systems that allow interactive visualization of the time-space prism. Figure 7.7 shows the plotting of one time-space route. In order to create the figure, the third dimension is given the value of time. The time used is the time relative to the time at which the visitor began his or her visit to the city.

The figure clearly presents the locations in which the visitor chooses to spend time. These locations are characterized by the straight vertical rising of the path, meaning that time is passing without movement. One of the advantages of computerized plotting is that the researcher is given the ability to view time-space data interactively and to change the view quickly and easily.

While plotting one route creates a comprehensive time-space diagram, when many routes are plotted at once the resulting diagram can be very difficult to interpret. Figure 7.8 is a time-space aquarium created using observations of forty tourists collected in Akko. These observations are all technically complete, no area suffered from poor GPS reception, and are also spatially spread out; routes of participants that did not leave the visitors' center were not included. Three main areas of activity can be recognized: the area circled on the right marks the visitors' center, the circle in the center shows the Turkish Bath House, and the left-hand circle shows the entrance to the Templars' Tunnel. Note that because all of the visitors began at the visitors' center, the column representing the time spent there is relatively compact while the column representing the time spent in the Templars' Tunnel (visited later) is much more spread out, representing the variance in the lengths of the visits.

Creating different views of the time-space aquarium using the computerized system makes it possible to find the best angle for viewing the figure and to recognize spatial trends within the data; however, it does not help in creating a typological division based on patterns of spatial behavior within

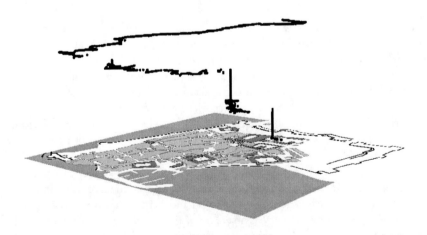

Figure 7.7 Plot of one time-space route in Old Akko.

Figure 7.8 Time-space aquarium presenting the activity of 40 tourists.

the data set. The use of three dimensions allows for a better understanding of the sequence of events throughout the visit but does not allow for any division or differentiation between different patterns of spatial behavior.

The new computation abilities that have become available within the past two decades have opened new possibilities in the plotting of space-time data. Kwan (2000; 2002B) was among the first to produce a "space-time aquarium" within a three-dimensional GIS environment using individual travel diary data. Kwan used the new computation capabilities to create both static and interactive space-time images. These images have improved researchers' capabilities in identifying characteristics of space-time patterns belonging to different population subgroups, and they have also improved researchers' abilities to identify common patterns. This approach to representing space-time activity is used later in this chapter to visualize the space-time data used for sequence alignment and also to visualize the aggregated results of the alignment.

One fundamental problem in space-time analysis is the aggregation of space-time paths to create generalized types composed of varied activities in order to identify patterns fashioned on a quantitative basis while taking into account the sequential element. Previous attempts with quantitative pattern aggregation methods, mainly by transport researchers (for examples see Pas 1983; Recker et al. 1983; Recker et al. 1987; Golob and McNally 1997; and Schlich and Axhausen 2003) did not manage to tackle the issue of the sequential element. Understanding the sequence of activities in space and time allows one to understand an additional integral dimension of activity and to recognize patterns that exist within this dimension. In his article "The Future of Geography," Nigel Thrift (2002, 293) foresaw the potential of sequencing methods used in genetics for geographic analysis. This chapter seeks to demonstrate the potential of sequence alignment methods as a means for aggregating space-time data while keeping the sequential dimension as part of the picture.

One form of analysis that has very promising potential for creating typologies of tourists based on their spatial behavior while taking into account the sequence of locations can be seen in sequence alignment methods (SAM). These methods, which have been used since the 1980s, were introduced to the social sciences by Abbot (1995) and Wilson (1998) and to the spatial sciences by Shoval and Isaacson (2007B) and Wilson (2008). These methods, which have developed with time and have been refined to more accurately compare sequences, have tremendous potential as a tool for creating typologies of tourists by analyzing their spatial activity. An explanation of the science behind sequence alignment methods and their development over time follows. For the interested reader, the section provides a deeper understanding of the methods' history and concepts.

SEQUENCE ALIGNMENT METHODS

Since the mid-twentieth century, scientists have grappled with the problem of how to analyze sequences quantitatively. The same challenge was faced by biochemists who wished to analyze protein sequences and social scientists attempting to analyze sequences of events (Abbott 1995). For years, the number of highly complex sequences and the complicated algorithms needed to analyze them combined to make any effective sequence analysis almost impossible. Then, in 1983, Sankoff and Kruskal, two mathematicians specializing in the field of applied mathematics, published a groundbreaking work in which they set forth a series of basic algorithms capable of efficiently analyzing complete sequences (Sankoff and Kruskal 1983). Their work, which formed the basis of most subsequent sequence analysis algorithms, proved key to the eventual breakthroughs made in DNA and protein analysis.

It took approximately ten years before the sequence analysis methods used with such great effect in the natural sciences filtered through to the social sciences. Abbott, a sociologist by profession, played a vital part in this process, demonstrating that sequence analysis methods could contribute not only to the natural sciences but to other fields as well (Abbott and Forest 1986; Abbott 1995). Abbott used sequencing algorithms to analyze the socioeconomic data he had amassed in the course of his investigations into the progress of musicians' careers. The first empirical study in the social sciences to employ sequence alignment, Abbott's work led to the development of "Optimize," a computer program that applies sequence analysis to socioeconomic data (Abbott and Hrycak 1990).

In the social sciences, sequence analysis methods have, to date, been used primarily to analyze sequences composed of a series of words or characters describing some aspect of human activity within a specific timeframe. Due to Abbott's pioneering work, sociology has led the field. (Examples of works include Halpin and Chan 1998; Blair-Loy 1999; and Stovel and Bolan 2004.) In the field of tourism research, Bargeman et al. (2002) aligned information describing vacation patterns. Transportation and time allocation research have benefited from work done by Wilson (2001), Joh et al. (2002), and Joh et al. (2005) in research that aligned patterns of activity. For a more elaborate review of the use of sequence alignment methods in the social sciences see Abbot and Tsay (2000).

Two types of analyses can result from a sequence analysis. First, there is the more commonly used product, utilized to construct groups based on their overall activity patterns. Programs using this kind of sequence analysis produce "trees," which in essence divide the sequences taxonomically. The second type of sequence analysis, less frequently employed, is used to detect patterns of behavior in some or all of the sequences scrutinized (Wilson et al. 1999).

How Sequence Alignment Works

Rather than creating groups based on the sequence of events within the object, traditional quantitative methods band similar objects together on the basis of some particular shared characteristic. Hence, the principal advantage of sequence analysis in the context of geographical studies, and specifically human geography, is that it allows one to explore systematically the sequential dimension of human spatial and temporal activity. In addition, sequence alignment methods differ from Markov chain analysis in two significant ways. The first is the outcome of the analysis: While sequence alignment methods can create taxonomic trees based on sequential differences between the sequences analyzed, Markov chains can calculate the probability of the next step based on the current location. The second difference is that sequence analysis analyzes the whole sequence where Markov chain analysis considers only the last location or a finite and unchanging number of locations prior to the current location.

As noted, traditional quantitative methods cannot expose the hidden patterns buried within sequences, which thus remain undetected (Wilson 1998). Unlike the more traditional methods of sequence comparison in which the distance between two sequences of activities is calculated by means of Euclidian geometry, such as Euclidean distance $[\sum_{i=1}^{n} (a_i - b_i)^2]^{1/2}$, city block distance $\sum_{i=1}^{n} |a_i - b_i|$, or Hamming distance (the number of positions in which corresponding elements are different), sequence analysis computes the distance between the two on a "biological" basis that will be explained in detail in the next section (Bargeman et al. 2002). The algorithm used in sequence alignment to measure the degree of similarity between two sequences utilizes three elementary operations: insertion, deletion (two operations, which are, on occasion, referred to singly as "indel"), and substitution (switching the places of two characters). By applying these three actions to one of the sequences, its string is made identical to the second string. The more operations needed to make the sequences identical, the greater the distance between the sequences. Distance and similarity are terms with opposite meanings; the greater the distance between sequences, the smaller the similarity. Other than that, the terms can be used interchangeably with the appropriate adjustment (Wilson 1998). Sequence alignment methods measure the degree of difference between two sequences in terms of element composition and sequence.

When used to measure two identical sequences (traditionally called source and target sequences) that have shifted, the Euclidian measurement will show that the sequences are very different when in fact the sequences of events were almost identical.

Example 1: Traditional Measuring of Distance (Comparison based on Hamming Distances):

Source sequence:	a b c d e
	x x x x x
Target sequence:	b c d e f

In this measurement, distance is measured by awarding 1 point for mismatched characters and no points (0) for characters that do match (a hit); accordingly, the higher the score, the greater the distance between two sequences. In Example 1, the distance is calculated as five (mismatched characters), which, due to the length of the string (five characters) is also the highest possible score. This, in turn, implies that the distance between the two sequences is likewise the greatest possible; however, it is clear that the two sequences are actually quite similar when considering the sequence of events. And therein lies the Euclidian method of analysis's basic flaw: It will calculate two sequences as very different even when they are virtually identical, thus making it almost impossible to detect similar sequences. This, in turn, makes it impossible to identify, let alone gain, any valid insight into the patterns of activity hidden within sequences of events.

Example 2: Comparison based on Biological Distance:

Source sequence:　　a b c d e
　　　　　　　　　　\ \ \ \
Target sequence:　　e a b c d

In this instance, the distance between two sequences is measured by the number of operations needed to produce identical sequences. In the case of the two sequences used in the previous example, it is possible, by inserting an "a" at the beginning of the target string and deleting the "f" at its end, to create two identical sequences. Since this process involved a mere two operations, the distance between the two sequences is now calculated as 2.

These simple examples demonstrate the advantage that sequence alignment methods have when attempting to recognize similar patterns that appear within activity sequences. In the first example, the distance measured did not show the similar pattern concealed within the sequence, a pattern that was revealed by measuring biological distances.

Clustering methods already in existence are able to aggregate observations based on evaluating the similarity of a series of different values. These methods calculate the similarity (or distance) between each pair of observations by summing the distances between each pair of variables based on a known function. This calculation of distances can be represented as follows: $D_{ij} = \sum_{i=1}^{k} d_{ij}$. A drawback of these methods is that, unlike sequence alignment methods, the observations must be of the same length (they must contain the same amount of variables) since they compare each pair of variables; thus sequences of different lengths must be made equal in order to enable a comparison.

The following example is a simple demonstration of how traditional clustering methods are unable to deal with the sequential dimension of data. The example consists of six observations. Observations 1–3 have sequences that are similar to one another and observations 4–6 are the same sequences reversed so that they contain the same characters in the opposite sequence. Figure 7.9A shows the sequences. The observations were analyzed using the JMP program's clustering options. The cluster method applied used the hierarchical method of clustering resulting in a dendogram, a taxonomic tree dividing the observations into different groups. The program analyzed each column separately and gave a score of resemblance (or distance) for each pair of values. Since our data are nominal, the distance function used was a value of 0 for 2 identical characters and a value of 2 for dissimilar characters. The distance between two observations was calculated as the sum of the distances between all matching values. The results of the cluster analysis are presented in a dendrogram (Figure 7.9B). The results of the clustering clustered sequences 2 and 4, demonstrating the inability of this method to decipher sequential similarity. In these examples we chose to use existing sequence alignment software that implements a multiple alignment (ClustalG).

	1	2	3	4	5	6	7	8
SEQ1	A	B	C	D	E	F	G	H
SEQ2	A	A	B	C	D	E	F	G
SEQ3	B	B	C	D	D	D	E	F
SEQ4	H	G	F	E	D	C	B	A
SEQ5	G	F	E	D	C	B	A	A
SEQ6	F	E	D	D	D	C	B	B

A

Figure 7.9A Example of sequences.

Since the introduction of sequence alignment techniques to the study of time-use by Wilson (1998), additional research has been done to improve the suitability of these techniques to the study of time-use. One example of the further development of this analysis technique is the development of a "multidimensional sequence alignment method" by Joh et al. (2002). This method allows for the alignment of more than one attribute (dimension) of activity. A second example of such an improvement is the development of a position-sensitive sequence alignment method (Joh et al. 2001). This method of sequence alignment is position-sensitive, meaning that the method is sensitive to the number of positions by which the sequential order of the target element is changed. Although not implemented in this study, both methods have potential for future space-time analysis.

ClustalG Software

Sankoff and Kruskal's algorithms and their various offshoots led to the development of a number of sequence alignment computer programs. As the majority of these programs were developed for use in biochemistry, their alphabet was limited to the twenty characters needed to denote

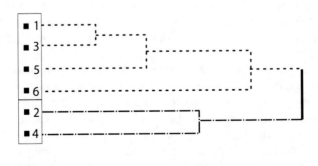

B

Figure 7.9B Dendrogram created using cluster analysis.

amino acids. In order to analyze socioeconomic data, a much larger alphabet was needed, which, representing as wide a range as possible of human activities (or, in our case, geographic locations), would allow researchers to tailor the programs in accordance with their specific needs. Several computer programs have either been written specifically for the social sciences or adapted from existing biochemistry programs. In this study, the ClustalG program, a general version of the ClustalX sequence alignment program used to analyze protein and nucleotide molecules, was opted for. The ClustalG program was adapted for general use by Andrew Harvey of the Department of Economics, St. Mary's University, Halifax; Julie Thompson of the Institute de Genetique et de Biologie Moleculaire et Cellulaire, Strasbourg; and Clarke Wilson of the Canada Mortgage and Housing Corporation. The Social Sciences and Humanities Research Council of Canada funded the project as part of its "Activity Settings, Sequencing, and Measurement of Time Allocation Patterns" project in 1998.

Eliminating the more exclusively biological features of ClustalX, and embracing an expanded alphabet representing myriad events or activities, the ClustalG program is eminently suited for social sciences research. One of the main changes made to the ClustalX program in order to make it more suitable for use in the social sciences was the extension of the alphabet accepted for alignment; this extension allows the use of words in addition to the twenty-six-letter alphabet. The use of multiple-letter words enables one to deal with almost endless classifications. This is significant because research in the social sciences tends to facilitate problems with broader classification than the limited number of amino acids that are relevant to biological research. Figure 7.10 shows the main alignment screen before and after executing an alignment.

ClustalG produces a complete multiple alignment in four phases. Phase one conducts a pairwise alignment; in this phase all pairs of sequences in the database are compared using the Smith-Waterman (Waterman 1995) dynamic programming algorithm for calculating similarity, producing a sequence-distance matrix. In phase two, a guide tree, which represents the similarity between sequences, is calculated from the distance matrix using Saitou and Nei's (1987) neighbor-joining method. This method creates a tree so that the sum of the tree's branches is minimal. Phase three builds a multiple alignment (see Wilson 1998). By following the guide tree created in the previous phase proceeding from the tips (leaves) to the root, for each node in the rooted tree the sequences of the left branch of the node are aligned with those on the right branch. This alignment is done using the Needlman-Wunsch (1970) global pairwise alignment algorithm. As in phase one, this alignment produces a sequence distance matrix. The fourth and final phase creates the final guide tree (Figure 7.11) using the same method that was used in phase two.

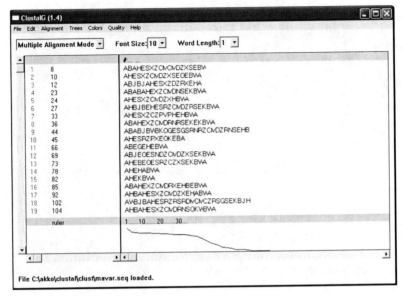

Figure 7.10A Alignment screen before creating alignment.

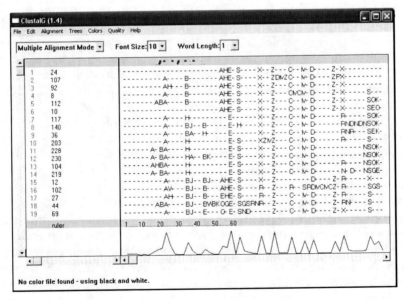

Figure 7.10B Alignment screen containing aligned sequences.

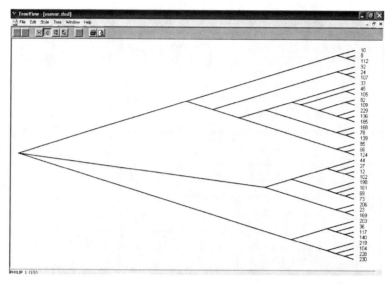

Figure 7.11 Final guide tree.

Applying Sequence Alignment Methods to Time-Space Data

This section explains the significance that sequence analysis methods have when applied to time-space data as well as the limitations these methods have when applied to spatial data. The first part of the section presents an example of a simple spatial alignment. The data for this example were invented in order to present a simple, easy to understand case for study. The alignment presented demonstrates the meaning that sequential methods have on spatial data. This demonstration is important in explaining the meaning that sequence analysis has on spatial data in a way that is easy to grasp. The combination of a complex analysis method and complex data creates a difficult media to understand; this demonstration aims at simplifying the data in order to enable understanding of the combination.

The second part of the section presents a series of experiments that were executed in order to uncover the implications that different scales in time have on the alignment. This problem is unique to the application of sequence alignment methods in the social sciences, where each character represents a unit that is determined by the researcher. In the natural sciences, in which each character represents a protein or DNA, the representation is predetermined for the researcher and he or she does not have to apply his or her judgment. The third part of the section presents two different alignments of the data collected in Old Akko, each of which addresses the temporal dimension differently. The fourth part of the section tests the beta release of a new version of "Clustal," a version that was adapted for analyzing spatial data and which takes into account the location of the activity as a separate parameter of the alignment.

An Example of a Simple Spatial Alignment

Figure 7.12A shows the sequences that were used to demonstrate a simple spatial alignment of a sequence of locations. The example uses six sequences describing motion through a linear space, and each character represents one time unit spent in an area. The first three sequences progress from A to F while the last three sequences have an identical route in the opposite direction. Figure 7.12B maps the linear space. An alignment of the sequences aligned the sequences around the "E" character, and was able to divide the data set into two groups, as can clearly be seen in the taxonomic trees (Figure 7.12C and Figure 7.12D).

Two trees are displayed here, a rooted tree and an unrooted tree. The trees distinctly show the two groups, but the unrooted tree is more difficult to interpret when a larger amount of sequences are aligned. Each of the groups consists of the sequences that moved through space in the same sequence of events. Figure 7.13 plots the results on a space-time diagram using the average length of time spent in each polygon for the time displayed. This example demonstrates the importance of the sequential component that lies within space-time data and its importance in achieving a fuller understanding of space-time activity. Sequence is an important dimension of spatial behavior and has a strong connection to the way people read and experience space in general and urban space in particular. Traditional methods would not be able to differentiate between the two types of sequences recognized by the method described above.

Sequence Analysis of Time-space Behavior of Visitors to Akko

In this experiment each character represents one minute, and so the temporal scale in this study had a much higher resolution than any previous sequence-alignment-based study. It is important to combine temporal and spatial resolutions that are compatible in order to retain the maximum amount of data without creating unnecessary "noise." The high temporal resolution of one minute used in this study was chosen due to the small geographical scale of the whole city and the small size of the polygons throughout the city; a lower resolution would have resulted in the loss of crucial data regarding the routes chosen by the visitors when moving from attraction to attraction.

Due to the small size of Akko's Old City, its area was divided into twenty-six polygons, with each polygon representing a single location. As mentioned earlier in the article, the ClustalG software allows the use of words in addition to the twenty-six-letter alphabet. The use of multiple-letter words enables one to deal with almost endless classifications. Therefore the decision to choose just twenty-six polygons was a matter of the small size of

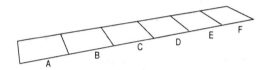

Sequence one -   AAABBBCCDEEEFF

Sequence two -   AAAABCDDDEEF

Sequence three - AABCCCDDDEEEFF

Sequence four - FFFEEEDDCCBBA

Sequence five - FFEEDDDDCCBBAA

Sequence six -   FEEEDDCCBBBAAA

Figure 7.12A Sequences representing motion in a linear space.

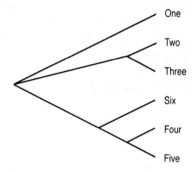

Figure 7.12B Alignment tree.

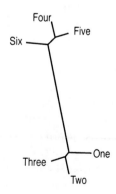

Figure 7.12C Diagram of the linear space.

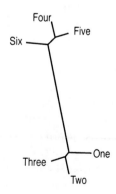

Figure 7.12D Unrooted alignment tree.

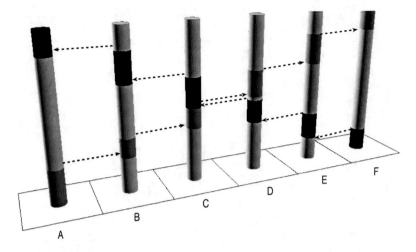

Figure 7.13 Time-space aquarium of motion through linear space.

the Old City (measuring no more than 125 acres) and of convenience, and not a result of a limitation of the software. In a case in which more locations are needed (e.g., in the case of a large metropolitan area), it is possible to define as many locations as needed by creating multiple-letter words. Thus, each tourist attraction was considered a separate location, the Old City's principal pathways were singled out and marked individually, and the marketplace (suq) was isolated from its surrounding area and divided into several sub-sections, each representing a different kind of market. As the locations themselves varied in size so did the polygons, each of which was allotted one of the twenty-six letters of the alphabet (Figure 7.14).

Using GIS, the visitors' GPS tracks were coded so that once every minute the system associated the letter assigned to the location in which the study's subjects found themselves at the time. The decision to use a high temporal resolution was made in order to ensure, among other things, that the study's temporal scale would match its micro-level geographical scale. It was also important to choose a resolution which, while not producing an unnecessarily long sequence, would still include all the relevant information found in the raw data that included the subject's location once every second.

One technical feature that needs to be considered is the function of the GPS receiver when it enters an area in which there is no ability to obtain signals and a loss of fix on the satellites is encountered. In such cases, the unit has logged the last known location obtained and continues to log it until it is able to record a new location. An interesting example of this was the behavior of the GPS receiver in the Templars' Tunnel; throughout the tunnel, where there is no GPS reception, the units logged the time spent in the tunnel as time spent at the entrance to the tunnel. This was taken into account when dividing the city into polygons.

Figure 7.14 Old Akko divided into polygons.

The final sequences of locations, in which each character represents one minute on the scale of time, were aligned using the ClustalG program. The alignment uses the default settings as set by ClustalG for the user-defined parameters. The alignment parameters that were used were as follows:

- Gap opening penalty, which means the penalty for an indel operation; in this alignment the cost of opening a new gap was set to 1.0.
- Gap extending penalty, the cost of every additional item in the gap, was set to 0.1.

The substitution matrix that defines the similarity between characters defined the score for identical characters and zero for all mismatches.

The alignment of the whole database produced a taxonomic guide tree (Figure 7.15A) that revealed the existence of three well-defined groups of visitors distinguished by (a) the number of locations visited and (b) the order in which they visited the locations. The first group included sixteen sequences; the second eighteen sequences; and the third group only six sequences. Each group was aligned separately in order to identify the highest scoring locations. At this point it was possible to continue and divide each group into subgroups according to the tree produced in the latter alignment, an option that was not applicable in this case due to the small size of each group. These locations, designated the alignment's "backbone," were marked down as characterizing the group as a whole (Figure 7.15B). The amount of time each of the individual groups spent in each polygon was determined by calculating the average time all of the group's members spent in said polygon. As noted, all of the visitors who participated in the study had purchased tickets to the Old City's principal archaeological sites including the underground Crusader City—the starting point of their tour of the city—and the Templars' Tunnel. As a result, the major spatial temporal differences between the three groups were the order in which they visited the city's remaining sites and the routes they took to reach them.

Figure 7.16 demonstrates the average time-space path of a group of visitors and is calculated taking into account the time spent in each polygon and the order in which they visited the different parts of the city.

Group 1: This group, taking the path (H) located immediately to the south of the exit from the underground Crusader City (A), walked towards the covered market (E), where they spent an average of fifteen minutes. They

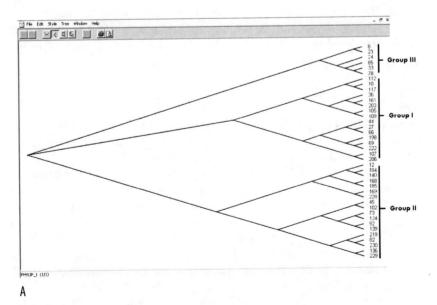

A

Figure 7.15A Taxonomic tree describing types of visitors in Old Akko.

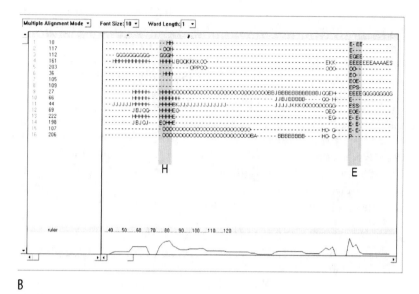

B

Figure 7.15B ClustalG screen after alignment.

then crossed the city to reach the Templars' Tunnel (Z). Having explored the tunnel they made their way to the Pisan Harbor (M) and the marina (D). They then walked back to the visitors' center via the covered market (suq) (E) and the open market (K), this time without stopping. This group did not visit the El-Jazar Mosque (J) or the bazaar just outside it (B).

Group 2: Having toured the underground Crusader City (A), this group headed straight to the El-Jazar Mosque (J) and adjacent bazaar (B). They then walked to the Templars' Tunnel (Z) by way of the suq (E). Having explored the tunnel, the group walked back to the visitors' center via the marina (D), the suq (E), and the bazaar outside of the El-Jazar Mosque (B).

Group 3: This group also took the path (H) immediately to the south of the exit from the underground Crusader City (A), but unlike the first group it bypassed the suq (E) and headed straight towards the Templars' Tunnel (Z). Having explored the tunnel, the visitors then walked to the marina (D) by way of the Pisan Harbor (M), where they stayed for an average of twelve minutes. The group then visited the bazaar alongside the El-Jazar Mosque (B). Having spent some twelve minutes wandering around the bazaar they returned to the visitors' center. Despite the fact that this group contained a mere six tracks, it was, nonetheless, useful in demonstrating how sequence alignment can be used profitably to identify and chronicle sequential differences. As can be seen, like Group 2, the members of this group also visited the bazaar outside of the El-Jazar Mosque, but at the end rather than the beginning of their track.

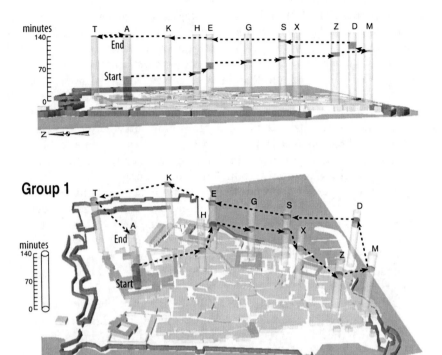

Figure 7.16 Typical space-time paths of visitors to Old Akko.

The next stage in analyzing the data was to try and discover whether there were any relationships between the three aforementioned temporal and spatial patterns, on the one hand, and the questionnaires which the subjects completed at the end of each of their excursions, containing information about themselves and their journey, on the other hand. Contingency Table Analysis was performed in order to check whether statistically significant relationships existed between the various categorical variables. Just one variable seemed to have significant effect (p = 0.0152) on the spatial activity; this was the "information" variable. It indicates whether people used the "Information Kiosk" at the visitors' center. Over 80 percent of the visitors belonging to Group 2 used the information kiosk while just 33 percent of the visitors belonging to Group 3 and 50 percent of the visitors belonging to Group 1 did so. However, the differences between the groups regarding this variable is an important finding, since in the research literature the information imparted to the visitor is considered to have tremendous impact over how the cultural tourism asset is ultimately used (Dann 1996) and significant influence on tourist movements in a local destination (Lew and McKercher 2006). Aside from this correlation, the statistical analysis failed to expose any other significant relationships between the variables. Due to the lack of tourist accommodation in the city, most visitors arriving to Akko intend to spend no more than a few hours there. This time is spent visiting Akko's premier sites, all of which are run by the Old Acre Development Company, including the visitors' center, the Crusaders' Halls, the Turkish Bath House, and the Templars' Tunnel. All of this leaves them with very little time, if any, to explore other areas of town, the result being only small differences between the three groups in terms of their members' personal characteristics and temporal-spatial activities.

However three distinctions between Groups 1 and 2 should be mentioned even though they were not found to be statistically significant. About 50 percent of the visitors in Group 1 included young children, compared to just below 20 percent in Group 2. None of the members of Group 1 had visited the Crusaders' Halls before, while in Group 2 more than 60 percent had visited this attraction previously; this could be the explanation for the shorter average visit of the members of this group at the site. Lastly, as indicated before, visitors belonging to Group 2 used maps, guidebooks, and other sources of information on Old Akko more extensively than did members belonging to Group 1.

The single most important spatial-temporal difference between the three groups discovered by the sequence alignment analysis proved to be the order in which they visited Akko's remaining sites and the routes they took to reach them. These findings, it should be noted, match those obtained by other similar studies, which also analyzed the spatial behavior of tourists in small historical areas (Dietvorst 1995; Shoval and Raveh 2004). For example, according to Keul and Küheberger (1997, 1011) personal preferences, goals, and plans, all of which play an important part in fixing most tourists' itineraries, are of little relevance when it comes to determining their activities on site. Indeed, tourists visiting Old Salzburg behave, as Keul and Küheberger have shown, in a manner remarkably similar to, even typical of, visitors exploring macro-scale, open-air museums or shoppers perusing the aisles of a supermarket—two highly regulated environments. The routes in such regimented environments are often tightly if discreetly controlled with people all too frequently subtly maneuvered into making particular space-time choices: when and from where to set off, and in what direction; when, where, and for how long to stop at specified locations; when to move on and to where. In small historic cities, which, as Keul and Küheberger observed, embody much the same environment, the result is an "ants-trail" structure of tourist pathways (1997, 1011). On the whole, and not surprisingly, it is much easier to predict the movement patterns of tourists visiting small, compact tourist destinations, which feature only a few attractions and possess a small transport system, than it is to predict those of tourists wandering around vast and complex urban destinations, which boast a great many attractions and a variety of tourist accommodations.

Akko, as a heritage tourist town, is strongly reminiscent of Edensor's (1998) Enclave Tourist Space. Enclave Tourist Spaces are virtually self-contained spaces, which, inasmuch as they are effectively cut off from the town's local population, "shield" tourists from such local "irritants" as noxious smells, unpleasant noises, unsightly slums, and bothersome beggars. The infrastructure of these spaces is, as a rule, specially tailored to meet tourists' expectations, as is the whole tourist experience on offer. This is far from a good thing in terms of the municipality's overall social, economic, and material well-being, which would clearly benefit if tourists were encouraged to spend both time and money in other parts of town as well. It is worth noting that virtually no visitors ventured into Akko's northeastern area, despite the fact that the area boasts a notable array of eighteenth-century fortifications, the very same fortifications which put a stumbling block in Napoleon's plan to conquer the town; and, which is perhaps even more surprising, despite the Old Acre Development Company's aggressive marketing of the area, and promotion of the so-called "Napoleon Route."

Testing a Spatial Sequence Alignment—"ClustalXY"

"ClustalXY" is a version of Clustal that was written explicitly for the alignment of spatial data. Users are required to supply two input files—the first lists activities and contains a key for identifying each individual; each activity points to its counterpart coordinate in the second file. This method allows for the alignment of both activity and geographic location and for the specification of a set of geographic locations for each participant in the research. A beta version of the program was made available to the authors for testing by Mr. Clarke Wilson, a member of the developing team.

This section presents the results of the testing of the beta version of the ClustalXY program using the data collected in Akko. An additional file including the geographic coordinates of the center of each polygon was created and saved in a format that could be read by the ClustalXY program and matched to the appearances of each polygon in the original file. Figure 7.17 shows the resulting alignment tree. This tree, like the tree created in the original alignment of the Akko data, includes three groups. In this alignment the two larger groups grew on account of the third group, which shrunk. Each sequence in the tree presented in Figure 7.17 is marked according to its location in the original alignment. The first large group consisting of sixteen sequences includes twelve sequences (75 percent) that belong to Group 1 in the original alignment. The second group consists of twenty sequences of which fifteen belong to Group 2 in the original alignment, again a matching of 75 percent. As expected, the results of the two alignments are very similar. It is important to note that when a group of sequences is aligned, the result that is presented is only one of the possible results; there is not one "correct" alignment for a set of sequences. This may lend to the differences between the two alignments.

This example did not fully utilize the new abilities that ClustalXY offers but rather repeated the same alignment done previously using a different method. The new program is especially useful for aligning data that are collected using activity diaries, data that include both activity and location, and data that describe everyday behavior. The program allows one to incorporate the differences in geographic location of different functional locations in a person's life, such as home and workplace, and to incorporate the differences in geographic distance into the alignment.

Unlike the natural sciences, in which the components that make up sequences are clearly defined, in spatial sciences the temporal and spatial resolutions are parameters that are manipulated by the researcher. This section demonstrated the importance of setting temporal and spatial resolutions that are optimal for the data's analysis. Paying attention to this point can eliminate alignments that do not fully benefit the data in question.

The combination of the dimensions of time and space creates sequences that are unique and complex, creating the need for a tool that is able to analyze both dimensions simultaneously. This section included an example of the creation of a typology of tourists based solely on the tourists' spatial behavior that has been made possible thanks to the high-resolution tracking data that can now be obtained. This is a unique kind of typology that is based on analyzing the sequential component of spatial data. In future research, it would be interesting to test the possibility of adding additional non-spatial characteristics to the spatial sequence data and using them as additional attributes to be taken into account when creating typologies of tourists.

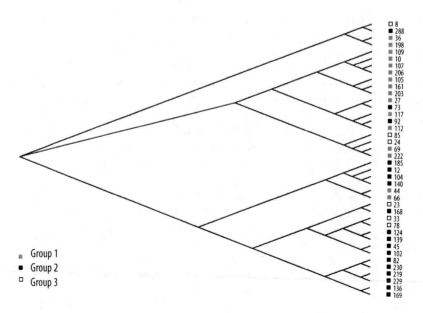

Figure 7.17 "ClustalXY" alignment tree.

This section presented one of the processes of applying sequence alignment methods. By aligning all of the sequences in a data set it is possible to divide the observations in the data set into groups of observations based on the sequence of locations (or events) within the data set. This method was shown to be effective in cases in which traditional clustering methods fail to divide sequences.

Another way in which sequence alignment methods can serve as data-mining tools is by enabling the matching of a known sequence to the sequences in a large spatial database. An example of this use can be the identification of a known sequence of events within the data set. For example, when analyzing data describing tourist spatial behavior, a fake sequence that is made up of a route that is recommended by an authority (a guidebook, map, or guide at the site) can be added to the data set and then aligned together with the original data. The resulting alignment will show which observations have sequences similar to the added sequence and therefore which visitors have followed the path recommended by the authority.

Additional ways to use sequence analysis are as a tool to assist in recognizing patterns in data describing spatial behavior. Recognizing patterns of normal behavior enables the recognition of unusual spatial behavior. Unusual spatial behavior can be interpreted differently when recognized within different populations. For instance, unusual behavior can indicate a cognitive impairment or a state of wandering if the subject is an elderly person or suspicious behavior if the subject is a convicted criminal.

Despite the usefulness of this method, it has several limitations. For example, it lacks a solid ability to assess the reliability of the alignment produced. Assessing reliability of an alignment in the social sciences and the sensitivity and weight that the alignment parameters have on the outcome of the analysis are important topics. Recently Wilson (2006) examined these questions and achieved partial answers based on examining data that were produced using Monte Carlo sequence generation.

An additional limitation of the sequence alignment software that was implemented in this test is the fact that geographical location is represented by a polygon that corresponds to one letter or a combination of letters. Therefore a process of generalization is conducted in order to translate the real-world locations into defined polygons. The spatial version of the Clustal sequence alignment software, which was tested in this study, should be able to overcome this limitation, since it is able to directly deal with and align true space-time coordinates and continuous space-time paths without any kind of spatial or temporal discretization. However, it still must be tested more extensively.

The study described in this chapter demonstrates the advantages of implementing sequence alignment in the context of a geographical investigation. The method offers an effective means for extracting sequence patterns from space-time data of human activities and by doing so may also present new ways of analyzing such data. It is equally worth noting that the relatively new and more accurate digital methods used in the study to collect spatial data (i.e., GPS) produced a database characterized by extremely high temporal and spatial resolutions. This reinforces Miller's prescient observation to the effect that recent developments in the field of Location Aware Technologies (LAT) and Location-based Services (LBS) could trigger an even wider resurgence in time-geographic studies (Miller 2005, 17).

CONCLUSION

This chapter presented the many different possibilities that time-space data present for the analysis and study of tourist spatial activity. These data have the potential of playing a central role in understanding tourists' mobility preferences and practices as well as movement patterns.

A central theme to which data that are collected using advanced technologies can contribute is the creation of typologies based on spatial behavior and the enrichment and deepening of the understanding of non-spatial typologies. At the time of writing, research in this area has just begun to emerge and we foresee that as tracking data become more widely available these theories will reach new levels of understanding and will be able to validate and strengthen existing theories.

A broad topic not discussed here is the use of data obtained by use of tracking technologies to feed simulations and agent-based models. These possibilities and their potential have been explored by Skov-Petersen (2005) in his paper titled "Feeding the Agents—Collecting Parameters for Agent-Based Models," where he explains what can be done in order to create agents that simulate human behavior. Tracking data can be used for the creation and/or validation of visitor movement simulations in destinations using agent-based models and other modeling techniques such as Markov chains to create data for agent-based models. In point of fact, the increasingly popular field of urban model-building (Batty 2005) could benefit greatly from the digitally based tracking technologies' ability to amass rock-solid descriptive data on spatial and temporal activities, using such data to both build and verify models. Not only would they provide these models with an exemplary fact-based database, but they would also help researchers to verify their models by seeing how these measures up to reality; for as White and Engelen (2000), noted: "In order to test such a model, it is necessary to compare the simulation results with the actual data" (396–397).

This kind of research, which is very technical and mathematical in nature, is in its first stages and can serve as a strong tool in assisting planners and policy makers to examine different scenarios before deciding on a chosen plan.

8 Understanding the Destination

Tourism in general, and in cities in particular, is a growing sector. Much urban tourism, researchers find, is concentrated in well-defined areas within the city. Leisure and cultural tourists are spending more of their time in the CTD (central tourist district), an area that usually includes a historic city center as well. Business travelers spend more of their time in the CBD (central business district) and in conference centers. Due to the increasing numbers of tourists, the spatial activities of tourists throughout different parts of urban centers are some of the forces that shape city centers as we know them today.

Tracking technologies present a great opportunity for the study of the impact that tourism has on urban centers and urban systems. As was seen earlier in the book, data collected using these technologies are more exact and can be gathered with greater ease and on larger scales in comparison with the time-space data that have been available until now.

The previous chapter put the tourist in the center of the discussion and presented the ways in which the analysis of time-space data collected using advanced tracking technologies can contribute to understanding the tourist's spatial activity throughout his or her visit to a destination. In this chapter, we look at time-space data collected using advanced tracking technologies but place the destination at the center of the discussion, enabling a greater understanding of how spatial activities of visitors generate different space throughout the location at different times and how visitors consume the destination itself.

The data presented in this chapter are aggregated figures that present the combined activity of many visitors in time-space throughout a destination. Such analysis can facilitate decisions such as where to set up new attractions and where to promote private-sector tourist services. This different angle opens many new points of view and questions that can now be addressed using high-resolution spatial data; these were virtually unobtainable using the traditional methods of data-collection on spatial activity.

The data obtained using tracking technologies can be analyzed in real time, creating virtual "radar" of the activity of visitors throughout a destination. New possibilities that arise include estimating the physical carrying capacity of attractions throughout the destination and of the destination itself; locating areas that remain out of the scope of the tourists' routes and that have unrealized potential that can be developed; and determining the effect that the time of day, weather, days of the week, and the seasons of the year have on the spatial consumption of the tourist destination.

Tourism, especially activities located within urban areas, which comprise a large percentage of the tourism industry, could greatly benefit from the kind of digital tracking methods that are able to trace pedestrian routes over long periods of time and, additionally, can do so both accurately and consistently. This is because the business, commercial, and leisure activities of most cities are largely concentrated in the city center, which is thus distinguished by high levels of pedestrian movement. This is true even in more developed urban economies, which have seen a move of business activity to the periphery in recent years; the town center in these remains at the heart of the town's social, cultural, and administrative life (Haklay et al. 2001, 343).

The cases discussed in this chapter use both GPS technology and data obtained from cellular networks. The GPS-technology data are collected by distributing GPS devices to visitors at a destination. This data can be downloaded from the devices when they are returned at the end of the visit and can also stream into the system in real time using cellular communications to transfer the data. Cellular network data can be processed close to real time as well or later on when online data are not available. The analysis of cellular network data presented in this chapter was performed by researchers using data obtained from the Estonian and Italian cellular networks.

AGGREGATIVE DATA FROM CELLULAR PHONES

Studying the aggregative patterns of tourists with data collected from cellular networks is possible due to the use of "roaming" phones in cellular networks. Cellular communication providers have agreements with providers in other countries in which the foreign providers are obligated (usually for an inflated fee) to give service to "roaming" end units from other networks.

These agreements mean that many tourists carry their cell phones with them and use them during their trips. Data are thereby created regarding the registration of devices on new networks—meaning that the phones are now located in a different country. For example, a phone that is usually registered in Vodafone Germany may be registered for a week on the network of an Estonian provider. In addition, the location of the phone (assuming the tourist carries the phone during his or her trip) can also easily be identified on a more specific geographic scale in some cases that will be described later in this chapter.

There are two main ways to analyze the aggregative data of cell phones' locations. The first is to analyze the statistics of the transceivers' activity for certain time periods, information that the operating companies routinely collect and use to manage the network. The other possibility is to detect the location and migration of a group of devices over a given period of time on the networks' different transceivers. This information is not collected routinely by the operating companies, making it more complicated to obtain; as will be seen later, this analysis could also entail issues of privacy, unlike the first approach.

In the first method, the number of devices that are associated with each transceiver of the network during a certain period of time is counted. This method has been called passive positioning (Ahas and Laineste 2006). In this method, no privacy issues arise, because specific cellular phones are not tracked; rather, a statistical analysis is conducted of the activity of the network's different transceivers. This method enables the researcher to sense the vitality of regions and cities and to analyze, for example, daily activity of tourists in the region and its cities. This is true if the type of analysis is undertaken only regarding phones "visiting" the network; otherwise, the analysis represents all networks subscribers, tourists and residents alike.

This approach also enables the researcher to integrate the data on human activity with other databases; for instance, integrating the data from weather radars allows the researcher to analyze the impact of weather on human activity. These data may also be used by the relevant public sector agencies for real-time control and management of large metropolitan areas in case of catastrophic events: for example, evacuations before major hurricanes or rescue operations after major earthquakes.

Janelle (2004) suggested that such data should be analyzed by the "Synoptic Analysis" approach, which relies on measures from data taken across a wide area simultaneously to identify space-time patterns and to forecast changes in relationships among these patterns over time. This analysis can provide a basis for automated real-time synoptic visualization of movement behavior in cities over certain periods of time (Janelle 2005).

The second method for analysis entails the detection of the location and migration of a group of devices over a given period of time on the networks' different transceivers. For example, the location of devices registered with the residents of a particular statistical district over a certain period of time can be analyzed, or, in the case of tourists, devices that clearly do not belong regularly to the network may be tracked. The aggregate data concerning human spatial activity can then be associated with certain types of spatially referenced data. Such activity could, for example, breathe new life into the census, making it possible to conduct a *Dynamic Census*. This approach may raise privacy issues; however, if sufficient aggregation levels are maintained, privacy can also be ensured.

When studying tourists' aggregative pattern in a destination, both methods can be employed. The simplest approach would be to include only "roaming" devices in the analysis. Another approach would entail having potential participants send text messages from their phones when they are willing to participate in a study. This can allow for the collection of additional background data from the participants or even asking the participants to respond to some questions. Using this approach allows for the inclusion of "local" cell-phone devices as well (i.e., it allows researchers to distribute local phones to tourist who are willing to participate) and does not limit the sample to roaming devices. This approach introduces a partial solution to privacy issues arising from the use of location data from cellular phones, since participants have consented to participate in the research.

Optimal Scale for Analysis

Cell phones have relatively low accuracy as a tracking method. Mateos and Fisher (2006) found in tests that the average accuracy was 3,625 meters (with a minimum of 500 meters and a maximum of 5,000 meters); other researchers found higher accuracies, between 150 meters and 1,500 meters (Shoval and Isaacson 2007a). The variation in results is mostly the product of the differences in the layout of the networks and the distance between cell towers.

This suggests that this technology is appropriate for studies of relatively low geographical resolution, for example on the metropolitan level, and not for high-resolution interurban analysis, where GPS is a better method for tracking subjects (Shoval and Isaacson 2006). In addition, it should be noted that the suggested methodology is better suited to research in cities and metropolitan regions, where the cellular network is denser than in suburban and rural areas.

Challenges for Implementation

Traffic monitoring using data from many cell phones simultaneously is already performed by private companies; the companies use aggregative cell phone data to monitor traffic in several cities worldwide (Richtel 2005). These companies usually sample phones that are connected to transceivers along highways and not whole networks. However, so far only a few efforts have been made to implement this concept in academic empirical research. The explanation lies, first, in the fact that this is a relatively new research area and, second, in the fact that that there are significant problems in acquiring the data.

The first effort to map human activity using aggregative data from cellular phones was made by a group from Tallinn, the capital of Estonia. Ahas and Mark (2005) conducted several studies with samples ranging from 30 to 117 subjects who carried cell phones for short periods of time. The first researchers to analyze a whole network of cellular phones on a large scale were researchers from the MIT Media Lab, headed by Carlo Ratti. They used cell-phone data to present the feasibility of using aggregative data for urban analysis in the metropolitan area of Milan (Ratti et al. 2006) and the city of Graz in Austria (Ratti et al. 2005).

There are currently several significant barriers that must be addressed before a widespread adoption of this methodology by researchers becomes possible.

1. Tracking technologies raise serious concerns regarding infringements on privacy (Fisher and Dobson 2003). They add a geographical dimension to the "surveillance society" (Lyon 2001) and enable better tracking of the "digital individual" (Curry 1997). However, in the first approach presented in this chapter, the data derive from statistics regarding activity in transceivers and not regarding the locations of the phones themselves—therefore there should not be any privacy considerations. The issue of privacy is relevant regarding the second approach only; however, the challenge can be reduced significantly by creating a "firewall" between the data supplier and the researcher in such a way that the individuals guarantee their own privacy. The solution could be similar to the solution employed in the case of censuses: detailed census records at the individual household level are not usually available to researchers. Data are available only in aggregated forms at a level of detail intended to prevent the ready association of demographic information with individual households.

2. Another barrier to the widespread implementation of this concept may be a lack of willingness on the part of commercial enterprises to share their proprietary data. These data are legally the property of the corporation that collects them and are not freely available. This means that such data are available only on terms and for purposes approved by and agreed upon by the owners. However, as we mentioned above, commercial companies and researchers have already received access and permission to use this kind of data.

3. Even in societies with a high percentage of cell phones, not everyone owns such a device, and even those who do do not always take the device everywhere with them. Thus, a full representation of the population may be a problem, although such issues must be addressed in any other research method as well.

In recent decades, governments in developed countries have invested considerable resources in launching satellites into space, thereby spurring the development of remote sensing and Global Positioning Systems. In the *physical* geographic domain, continued rapid developments in remote sensing have dramatically increased the availability of data describing earth surface processes such as climate, changes in land cover and land use, deforestation, and urbanization (Longley 2002). Each of these new sources of physical geographic data is related to aspects of human spatial activity, but none of them can be thought of as materially augmenting more traditional social scientific data that describe the social characteristics of individuals and groups.

At the same time, the private sector has established a global infrastructure for the operation of cellular phones in many parts of the world. The development of this infrastructure could augment the possibilities for research of spatial aspects in the social sciences in general and tourism in particular.

Examples of Studies Using Data From Cellular Phone Systems

One of the main difficulties in using data from cellular networks, as mentioned earlier, is the actual attainment of such data from cellular network operators. Two examples of such studies exist and will be accounted for briefly in this chapter. Cellular operators, which are privately owned companies, do not always have incentive to cooperate with researchers. This makes close contact with the operators an essential prerequisite to this kind of research.

This is the case in Estonia, where researchers in the Department of Geography at Tartu University have achieved a very close working relationship that allows them to receive data from the cellular network regularly. Another research group that was able to access cellular phone network information is the SENSEable City Laboratory, located at the Massachusetts Institute of Technology and headed by Carlo Ratti.

Estonia

Two papers published by an Estonian group of researchers headed by Rein Ahas (Ahas et al. 2007, Ahas et al. 2008) used cellular network information to gain insight into the activity of tourists in Estonia. In the first paper, published in 2007, cellular phone data was used to establish spatial patterns of tourists in different seasons.

The study found that the visitors to Estonia are located in different parts of the country in different seasons. The summer months benefit from large numbers of tourists who mainly spend their time in the Baltic coastal regions; the winter months have a higher percentage of tourists in the northeastern region of Estonia, possibly because of the famous spa hotels located there and the ice fishing on Lake Peipsi in the region.

Generally, the findings strengthened what was already known about tourists in Estonia from other sources, primarily from accounts given by hotels regarding the numbers of tourists using their services. One finding that was surprising and that had not been recorded previously was the flow of Latvian fishermen to Lake Peipsi in the winter months. These fishermen are not accounted for in other data sets, since they can access the area freely, are not registered at the border, and usually do not use any regular tourist services since they do not stay overnight. This population can be a source of growth to the Estonian tourism industry and had been mostly unknown up until this study due to the fact that they did not stay overnight and were not registered at the border as Latvia and Estonia have joined the EU.

The second paper published by the Estonian group (Ahas et al. 2008) focused on the flows of tourist movement as can be observed by the records describing the roaming phones in the cellular phone network's database (Figure 8.1). The article also attempted to examine the use of cellular phone data as a tool for tourism research by cross-referencing the analysis of roaming phones on the cellular phone network and traditional information regarding accommodations, finding a high correlation between the two sources (Figure 8.2).

Both of these studies were carried out using passive positioning. Using passive positioning has the advantage of allowing the analysis of large amounts of data relatively inexpensively but it also has some drawbacks. One main drawback of passive positioning is that the location of a cellular phone is only recorded when the phone has call activity—incoming or outgoing phone calls or text messages. While incoming calls can mostly be considered random since the person receiving the call has little control over when the call will come in, this is not the case with outgoing calls. Outgoing calls are not distributed randomly over space; rather, there are locations and times when they are more likely to take place, such as when people are waiting in a traffic jam or find themselves with free time. In addition, the high costs of using a cellular phone when abroad decrease the number of calls made or received (some people tend to filter incoming calls by enabling a mechanism that requires the caller to dial a code), since the subscriber generally pays for incoming calls as well when abroad. Therefore, data obtained passively have a bias toward those locations in which people tend to generate calls on their cell phones more often and towards the kind of tourist who makes more calls (e.g., business travelers, whose companies may pay the bill).

Habits in the use of mobile phones differ from country to country and culture to culture; this can lead to an indirect relation between the representations of visitors in the cellular network to their actual presence. Pricing for roaming varies between countries and between different carriers in the same countries and can be quite costly, leading to different patterns of usage. Differences in technology can also be a barrier to inclusion in the roaming phone database. When traveling to a country where the cellular phone technology used is different than that of the home country, it is not possible to use the same end unit as a roaming phone. This leaves out certain tourists, for example American tourists in Europe (who do not have a device that operates on a GSM network), and can create a bias in the roaming phone database.

Rome

A collaboration between Telecom Italia (TI), the cellular communications provider that serves 40 percent of the Italian network, and MIT SENSEable

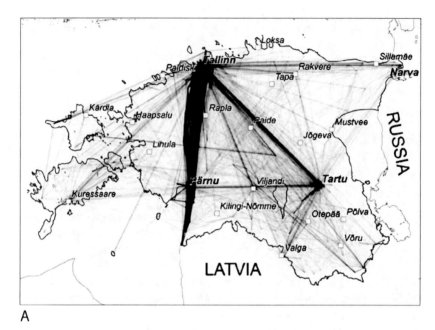

A

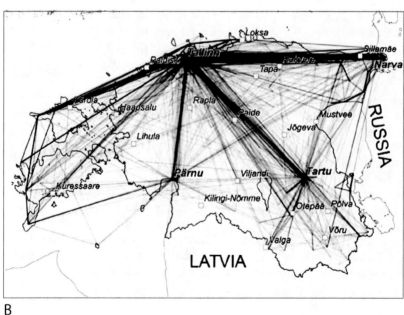

B

Figure 8.1 Linear movement corridors of (A) Latvians and (B) Russians in Estonia during the Midsummer's Day holidays (June 22–25, 2004). Source: Ahas et al. 2008.

Figure 8.2 The correlation of monthly sums of accommodation and call activities by counties. Source: Ahas et al. 2008.

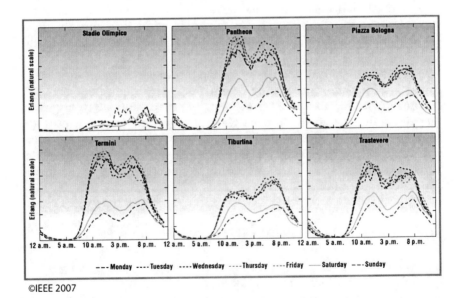

©IEEE 2007

Figure 8.3 Erlang data for the six locations by day of week. Source: Reades et al. 2007.

City Laboratory led to an innovative presentation of cellular network data at the Venice Biennale in 2006 (Reades et al. 2007, Calabrese and Ratti 2006). The installation presented both real-time and historic data regarding mobile phone usage in central Rome. The main source of data used was the Erlang data provided by TI. This information, collected at the transceiver level, is a measurement of the network bandwidth usage at a given time.

These data were measured at six different locations and presented on a graph on each day of the week (see Figure 8.3). The resulting graphs show distinct temporal patterns of usage at each location and also illustrate the difference between the usage of cellular phones during the week and on weekends. These graphs demonstrate that, as discussed above, cell phone usage is not randomly located throughout the city and cell phones are used differently in different locations.

Although not geared toward understanding tourism, the analysis and representations carried out using the cellular data in Rome enable the depiction of the urban environment into which thousands of tourists flow in ways that were not possible before this source of data was made available. Sensing the pulse of the city center, the life in urban areas, and the vitality of the tourists within these areas sheds new light on the two-sided relationship between urban spaces and the people who shape them.

Figure 8.4 is a visualization of Erlang data during Madonna's controversial concert (in which she staged her crucifixion, ignoring the Vatican's protests). The concert was attended by over seventy thousand people, creating a very visual peak in cellular phone usage in the Stadio Olimpico, where the concert took place. Visualizations of this nature are visually stimulating and easy to grasp intuitively but are "quite difficult to interpret rigorously, and they provide little insight into local-area dynamics without additional processing" (Reades 2007, 31). However, there is no doubt that this project was an important achievement in the field of research using cellular phone data; it has paved the way for future research by demonstrating the extent to which information from cellular phone networks can be used to benefit the understanding of the spatial dynamics of the phones' users.

AGGREGATIVE DATA OBTAINED FROM GPS DEVICES

In this section we present studies that were carried out using GPS receivers in five different locations: PortAventura amusement park and the Mini Israel theme park (two enclosed outdoor environments), the Old City of Akko (a small historic city), and the city of Rouen, France (with its historic city at the center of a five-hundred-thousand-person metropolis). The fifth case presented is Hong Kong, which serves to represent large multifunctional cities and large-scale regions.

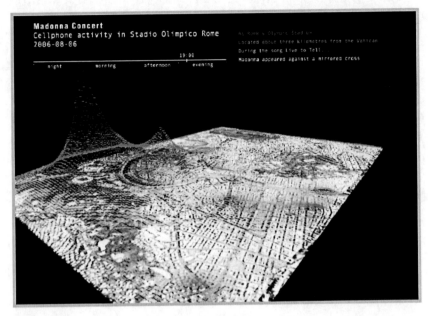

Madonna Concert
Cellphone activity in Stadio Olimpico Rome
2006-08-06

©IEEE 2007

Figure 8.4 A three-dimensional plot of telecommunications activity during a Madonna concert in Rome. Source: Reades et al. 2007.

PortAventura

An example of an exploration of aggregative time-space activities of visitors within a destination is a study that was conducted in the PortAventura amusement park. The project was a joint venture of the Universitat Rovira i Virgili School of Tourism and Leisure, the Hebrew University of Jerusalem, and the theme park itself.

As we saw earlier, PortAventura theme park is located in Catalonia, Spain, next to the holiday resort of Salou, approximately one hour's ride from Barcelona. Though identified with Universal Studios, the park is owned and operated by the Caixa banking group, which bought Universal Studios' shares in the park in 2004. Adjacent to the park are three hotels operated by the park and many others that are privately owned. In recent years, the park has exceeded four million visitors annually.

Park Structure

PortAventura is divided into five thematic areas. Each area represents a different geographic region and is designed according to the landscape and cultural characteristics that distinguish that location. The thematic areas are: the Mediterranean, Polynesia, China, Mexico, and the Far West. The Mediterranean area is located at the park entrance, while the other four areas are arranged in a circle. This means that a visitor who arrives at the park has to cross the Mediterranean section and then decide whether to circle the park from the right, starting from Polynesia, or from the left, starting from the Far West (see Figure 8.5).

Fieldwork and Sampling

As we saw in Chapter 6, the fieldwork for the PortAventura study was conducted in two phases of one consecutive week each. The first phase took place during the spring of 2008 and the second phase during the summer of 2008. The sample was restricted to families with young children. Of the 288 families who took part in the study, 277 families were included in the final analysis (96 percent). Three different types of data were collected for each family using three different data-collection methods (see Figure 8.6):

1) Visitors' socio-demographic and personal data were collected by park employees at the park entrance using a conventional questionnaire.
2) Time-space data were obtained from the GPS devices, which were set to sample the location of the visitors every ten seconds.
3) Data regarding the visitors' decision-making were collected using designated software. When visitors returned from the day at the park, GPS data were automatically processed, and a table specifying the park sites visited by the families was produced (see Table 7.1 in the previous chapter). This table allowed the interviewers to question the visitors as to their motivations and decision-making processes during their visits. (This application of real-time analysis of time-space data was discussed at length in Chapter 7.)

The sample of visitors who were asked to participate in the research was restricted to families with young children. This choice was motivated by the fact that a small group was needed to make the sampling statistically significant, and that this group was of interest to the park management.

As we saw earlier, there were a total fourteen days of sampling in two rounds of one week each, in April and July 2008. Twenty GPS loggers were used (for technical details regarding the receivers see Chapter 6). A total of 288 GPS tracks were recorded and 254 interviews were conducted by the park's staff. The missing interviews were primarily due to the fact that the park did not have staff available to interview all of the participants in a timely fashion. Before the study began, the park's management hoped to interview approximately one quarter of the participants; the results far exceeded everyone's expectations.

Information collected from eleven participants was excluded for the following reasons: Four participants carried devices that turned off during

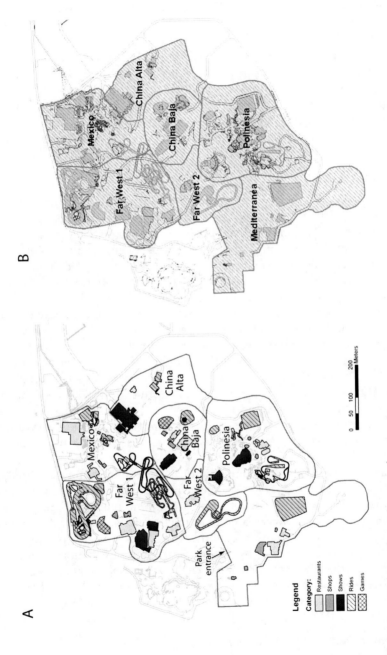

Figure 8.5 Map of PortAventura theme park.

their visits; six participants had devices that did not function properly and thus the spatial data were incomplete; and one participant who participated was mistakenly included (he did not meet the inclusion criteria). The size of the final sample used for analysis included 277 participants, or 96 percent of the total sample.

The external conditions within the park differed between the two phases of research. The park was open for nine hours a day in April (10:00 a.m.– 7:00 p.m.) and for thirteen hours a day in July (10:00 a.m.–11:00 p.m.). In April there were low temperatures with light showers from time to time and in July there were high temperatures with one day of extreme weather conditions that included heavy rain in the afternoon and evening.

Time Distribution

Figure 8.7 details the average amount of time the participants spent in various parts of the park. The first thing that the reader notices when looking at the diagram is the striking imbalance in the integrated amount of time spent in each zone and the relation between the time spent in the different areas. The subjects spent the most time during the first pilot, in the Far West section of the park (Far West2) The extra time that the visitors had in the second phase due to the longer opening times was mostly spent in Polynesia. This is a very interesting finding. One would think that the time that was added to the visitor's time budget as a result of longer hours of operation would either be divided evenly over space or divided over space in proportion to the popularity of the zone. Knowing that most of the additional time allocated to visitors is spent in one zone has great importance for the park's management, which must allocate employees throughout the park and needs an understanding of how to deploy its staff in the most efficient way. These findings are very useful when considering the operation costs incurred as a result of the longer operation hours.

If a city were studied in a similar manner, the results obtained by analyzing the way in which space is consumed could help the city's tourist authorities formulate a more reasonable tourist planning policy: A policy aimed at managing tourist flows in a more rational manner, a policy deliberately designed to relieve the burden from the town's more congested areas, both at set times and in general, by, among other things, encouraging tourists to explore other, less crowded sites. The result of such a policy would be a more equally distributed pattern of tourist temporal and spatial activity, one that could benefit both tourists and the town as a whole.

Temporal Cycles of a Destination

Figure 8.8 shows where the participants spent their time throughout the cycle of a day at the PortAventura theme park in each of the two phases of the study. The longer opening hours in the second phase can be seen on

The data collection procedure

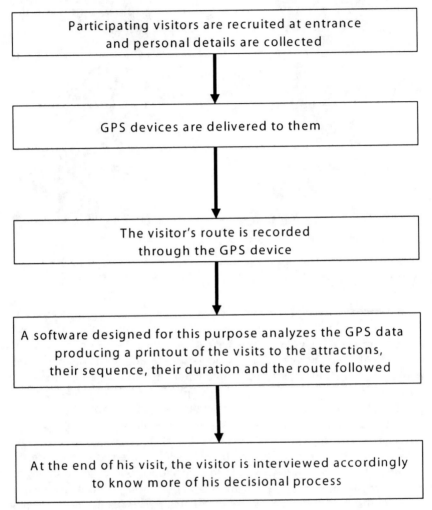

Participating visitors are recruited at entrance
and personal details are collected

GPS devices are delivered to them

The visitor's route is recorded
through the GPS device

A software designed for this purpose analyzes the GPS data
producing a printout of the visits to the attractions,
their sequence, their duration and the route followed

At the end of his visit, the visitor is interviewed accordingly
to know more of his decisional process

Figure 8.6 Data collection procedure.

the x-axis. These graphs were created using the map of attractions produced by the park's management. The map contains five different types of attractions: rides, shops, shows, restaurants, and games. Each attraction was represented using a polygon shaped to include the waiting line for the attraction and the area of the attraction itself. This is important because many of the rides are very fast and locating the participant riding on a roller coaster can be difficult if the participant does not spend time waiting on line.

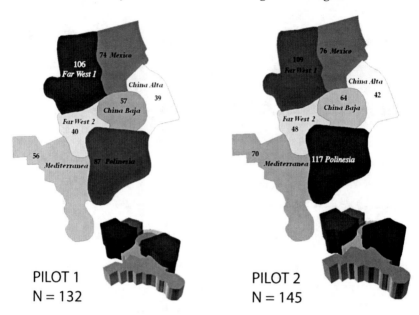

Figure 8.7 Average length of stay (min.) of all visitors in the park's thematic areas.

The total volume of activity was at its highest shortly after the park opened and all of the GPS devices were distributed. As the day progressed, the total volume of activity diminished as people slowly left the park. Both figures show a similar pattern: People enter the park and rush to get on rides, resulting in a peak in the ride graph. As the day progresses, people spend less and less time on the rides. The restaurant graph shows a clear peak at lunchtime during both phases and another much smaller increase at dinnertime during the second phase when the park was open until midnight and people ate their dinner there. Peaks in the show graph can be explained by the schedule that most of the shows follow (some shows are open all of the time and people walk in and out as they please). Both games and rides display very low volumes of activity. This may have nothing to do with the actual spatial activity of the participants; it may be linked more with the limitations of the GPS technology, which had difficulty locating participants within the very small store and game polygons. In addition, the stores were mostly located within built structures, making for a more challenging environment for the GPS.

Mini Israel Miniature Park

Mini Israel is a park located in Israel, midway between the capital, Jerusalem, and the economic center, Tel Aviv. The park hosts hundreds of

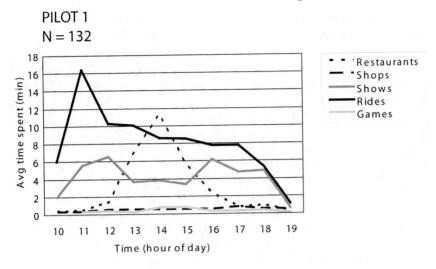

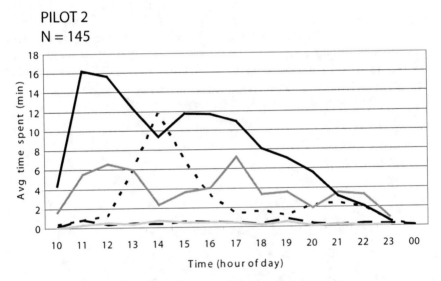

Figure 8.8 Visitors' time budget by hour of the day, PortAventura.

miniature models of key locations in Israel. The layout of the models in the park does not reflect the physical structure of the country; rather, the park is shaped like a Star of David.

A study of the time-space activity of visitors to the park was carried out in the summer of 2006. During the study, visitors to the park were approached and asked to carry GPS units with them throughout their visit. The park is outdoors and has very few buildings; it is therefore an ideal location for using GPS technology.

The data obtained in the study were used to create a typology of the use of the parks by visitors. Four categories of the areas in the park were created based on how those areas were used by visitors. This analysis is presented as a demonstration of a practical tool for understanding the spatial activity within a destination and managing the way visitors flow through an attraction.

To carry out this analysis, the area of the park was divided onto a raster (a grid); each cell was sized at 2 x 2 meters. The number of visitors that passed through each cell in the grid was counted. The cells in the grid were then classified into two categories: high-traffic cells and low-traffic cells (Figure 8.9A; high-traffic areas are dark and low-traffic areas are light). At the same time, another raster was calculated. In this raster, the average length of a visit was calculated for each cell. As with the first raster, the results were divided into two categories, cells with long average stays and cells with shorter average stays (figure 8.9B; cells with high average stays are dark and cells with low average stays areas are light).

Combining both categories using the criteria presented in the table below resulted in a grid with four categories. The categories are displayed in Figure 8.10. Each category explains the way that the visitors to the park used its space. Some areas in the park are used as corridors through which people pass but do not spend time; other areas serve as basins that channel the flow of people into them.

A

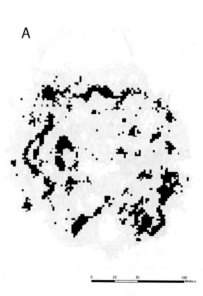

Figure 8.9A Classification of cells into high-traffic and low-traffic cells according to the amount of people that passed through each cell.

B

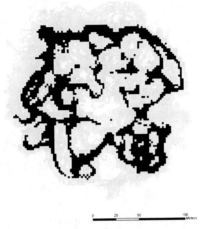

Figure 8.9B Classification of cells into high average and low average according to length of stay.

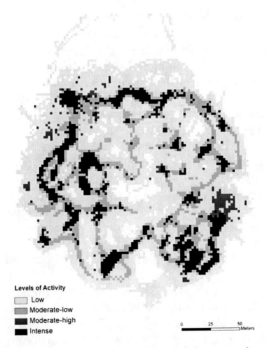

Levels of Activity
- Low
- Moderate-low
- Moderate-high
- Intense

Figure 8.10 Classification that reflects the usage of the space throughout the Mini Israel park.

Table 8.1 Cell Classification, Defining Activity in Each Cell

	Low-traffic Area	*High-traffic Area*
Short Average Stay in Cell	Low activity, low traffic, and short stays.	Moderate–low levels of activity, high traffic, short stays.
Long Average Stay in Cell	Moderate–high levels of activity, low traffic, long stays.	Intense activity, high traffic, and long stays.

The Old City of Akko

Background information about Akko and the description of the field there can be found in Chapter 6 and will not be repeated here.

Visitors' Impact and Management Implications

The GPS devices carried by the experiment's subjects throughout their entire tour of Akko's Old City were used to collect data on the subjects' locations. They did so by registering the precise location and exact time each location was logged. This spatial-temporal information was recorded at an extremely high level of intensity: one location per second. This meant that if, for one reason or another, the GPS satellite signal was blocked, as is often the case in dense urban environments, it was possible to reestablish a connection, and thus obtain a reading, the moment the device acquired a direct line of sight to the satellite system; this in turn meant that any breaks in the track sequences were reduced to an absolute minimum. The data obtained were used to analyze the spatial and temporal behavior of each visitor as was explained in detail in the previous chapter.

Here we wish to present the potential of the aggregate analysis that was done for the entire body of subjects. This analytical approach looks at the subjects' effect on the city or how the city is "consumed." Based upon aggregate data rather than single observations or clusters of observations, such a city-centered analysis can be used to indicate which are the more popular sites and neighborhoods in a town and which tend to be neglected; or, alternatively, which of the town's routes are well worn and which remain virtually unused.

The data obtained using the GPS devices were analyzed in aggregate. Such an analysis, which ignores the individual visitor, serves to reveal how the urban space is exploited or "consumed" by all tourists. In our case, the high-resolution data provided by GPS were used to create a "pixilated" map of Akko, which highlighted just how the town's urban space was consumed. The spatial consumption was measured by percentage of time spent

Figure 8.11 Tourist activity in Old Akko—two-dimensional.

in the different locales plus the intensity of activity per cell of a size of 10 meters by 10 meters, which is two times above the average accuracy of the 5 meter accuracy of the GPS units used in this study (see Figure 8.11).

Pixelating the Town

Another means of examining temporal and spatial behavior in aggregate, one which similarly exploits the advantages of the GPS system in terms of providing accurate and high-resolution data, consists of dividing the town's urban space into squares and counting the total number of signals picked up by the GPS receivers per square. In the case of Akko, the Old City was divided into squares measuring 10 meters by 10 meters. Obviously, the size of the grid's squares depends on the size of the urban space studied, with larger-scale studies requiring larger squares and smaller ones smaller squares.

Before it is possible to begin the discussion on three-dimensional visualization, it is important to note that in this chapter the three-dimensional visualizations are actually pseudo-three-dimensional figures. The figures are rendered in a way that gives the illusion of a third dimension but, since they were created on a computer screen and printed on paper, this remains an illusion only.

Adding a third dimension opens new possibilities for plotting time-space data. Before discussing the main contribution that the third dimension allows for (i.e., for representing time, as discussed in the literature review) we will present a group of figures that use a third dimension to represent factors other than the time that has passed. The third dimension, along with the color that was used in the two-dimensional figures, can help the researcher to understand the different amounts of time that each unit represents. Adding a horizontal dimension helps the researcher to understand the difference between the different quantities; this is made possible due to the fact that height is a scalable dimension while color scales are much more difficult and less exact for the human eye to interpret. This technique of plotting is very effective in studying the general impact that spatial behavior has on the location being studied.

This technique allows the researcher to distinguish between "Hot Spots," locations that are well-exposed to the visitor; "Not Spots," locations that do not exist for the visitor and are not visited at all; and transition areas through which the visitor passes but does not stop to spend time. This technique analyzes all of the time spent in the city and therefore is useful for learning about the city; however, it is not effective as a tool for shedding light on individual tourists.

Figures 8.11 and 8.12 depict which areas in town boast high levels of concentrated tourist activity and which plainly suffer from a dearth of tourists, darker colors mean more intensive levels of activity. Indeed, viewed as whole, the map exposes a marked spatial imbalance between the town's sites, an imbalance rooted in the way Akko's tourist industry developed over the years. Of all the columns in Figure 8.12C, the most prominent column is located in the area containing the visitors' center, the underground Crusaders' Halls, and the Turkish Bath House.

The figure also reveals which of the possible routes linking Akko's various centers of activity are the most commonly used. Apparently, most visitors to Akko tend to move in a southerly direction: setting off from the visitors' center and ending up, by way of the local market, at the Templars' Tunnel and the restaurants alongside the marina. Once the subjects who only visited the Crusaders' Halls are weeded out from those who explored the town's other sites as well, the resulting diagrams' topography, not surprisingly, changed. Thus, Figure 8.12A based solely on the sequences of those subjects who limited their visit to a tour of the visitors' center, Crusaders' Halls, and Turkish Bath House complex, contains a single concentration of fairly tall columns; while the topography of Figure 8.12B, based on the remaining sequences, is, predictably, much more diffuse.

A

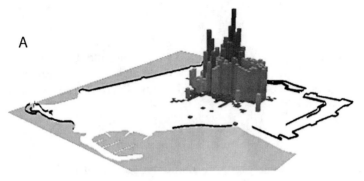

Figure 8.12A Tourist activity in Old Akko—three-dimensional: Visitors who limited their visit to a tour of the visitors' center, Crusaders' Hall, and Turkish Bath House complex.

B

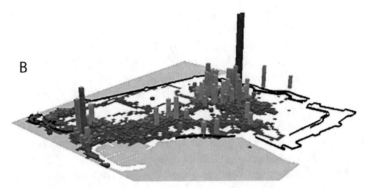

Figure 8.12B Tourist activity in Old Akko—three-dimensional: Visitors who ventured out of the limited area described above.

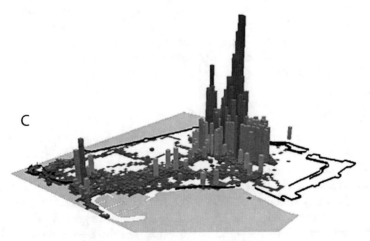

Figure 8.12C Tourist activity in Old Akko—three-dimensional: The full sample of visitors to Old Akko.

Tourist flows in Akko's Old City are dispersed unevenly throughout the town's various locales, with a great many visitors venturing no further than the visitors' center, the Crusaders' Halls, the Turkish Bath House complex, and the Templars Tunnel—all sites run by the Old Acre Development Company. These findings, it should be noted, match those obtained by other similar studies, which also analyzed the spatial activity of tourists in small historical areas.

Rouen

Rouen is the historic capital of Normandy in northwest France. Located on the Seine River, this city dates back over one thousand years and consists of a historic city center that is similar to Akko in that it is made up of stone buildings and small walkways. Unlike Old Akko, which is mainly a tourist city with a small local population, Rouen's historic center serves as the main city in a metropolis of over half a million people and within the historic Old City there are many modern shops that serve the local population.

In a study carried out as part of the "Spatial Metro Project: Improving City Centres for Pedestrians," which studies the pedestrian use of five European cities— Norwich, Rouen, Koblenz, Bristol and Biel/Bienne— Stefan van der Spek from the Technical University of Delft gave out GPS units to visitors to the historic city center (Spek 2008B). The GPS devices were distributed and collected at two parking facilities during one week. The time-space data were collected in combination with a questionnaire consisting of social-geographical data.

Figure 8.13 shows one of the results of the tracking conducted in the city of Rouen. Figure 8.13A plots tourists visiting the city for the first time while Figure 8.13B plots the tracks of regular visitors to the city center. An examination of the figures shows the different ways in which the two types of visitors consumed the urban space. The regular visitors spread out through the whole area, walking through large portions of the area adjacent to the parking lot, while the first-time visitors used the main routes and consumed a much smaller part of the city. This study demonstrates how the information barrier can limit spatial activity and create patterns of different natures.

Hong Kong

An example of a preliminary analysis of aggregative use of space in a large scale multifunctional world city of about seven million inhabitants can be seen in a small sample of a study in Hong Kong. The data presented belong to a two-year project (2008–2010) that, at the time of writing, had recently begun. The study is being conducted as a partnership between Robert (Bob) McKercher of the School of Hotel and Tourism Management at the Hong Kong Polytechnic University and Noam Shoval of the Department of Geography at the Hebrew University of Jerusalem.

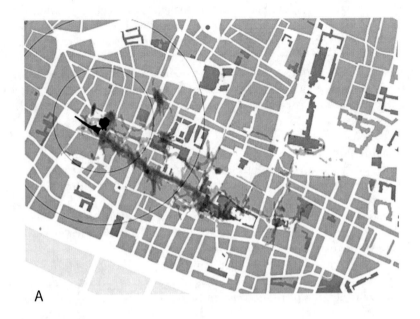

Figure 8.13A Visitors to Rouen—first visit.

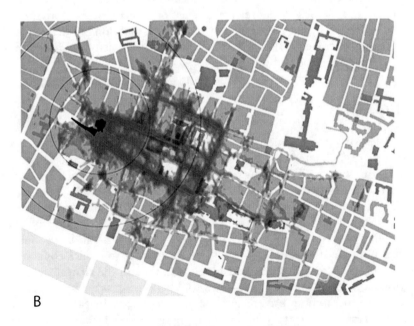

Figure 8.13B Regular visitors to Rouen. Source: Spek 2008B.

The two parts of Figure 8.14 presents the aggregate activity of first-time visitors to Hong Kong (n=77) and of the repeat visitors (n=40) who were tracked with a GPS device in Hong Kong and its adjacent territories and islands. The tourists were sampled at the lobby of the Harbour Metropolis Hotel in Kowloon and were requested to carry a GPS device for the day. The central part of Hong Kong (including Kowloon and Hong Kong Island) was divided into a raster of 200 meters by 200 meters. The accumulative time spent by all the tourists in each pixel is represented in the figure in a three-dimensional view.

When looking at the two figures, it is clear that the highest peak in the figure is the Harbor Metropolis Hotel (A) where the tourists were sampled. This is a result of tourists waiting for the shuttle bus or returning to the hotel to rest in the middle of the day. Extensive time was spent in the shopping district in south Kowloon (B) around Nathan road. However we can see that first-time visitors tend to be less represented in the various markets around Mongkok (C) in Kowloon.

On Hong Kong Island it is interesting that the tourists did not spend significant time in any of the built areas in the northern part of the island (D) aside in the area of the ferries landings, but this may be a bias based on the place of sampling that is located in Kowloon. This hypothesis will, we hope, be validated in the future, as the plan is to conduct some of the sampling in hotels on Hong Kong Island.

Some repeat visitors did, however, explore the district of Causeway Bay (E). As expected, there is a significant concentration of activity of first-time visitors in the Victoria Peak area (F) and less of repeat visitors. Both groups visited Stanley (G), but it is somehow surprising that none of them visited the area of Aberdeen (H) though repeat visitors did visit the Ocean Park theme park that is located nearby (I).

CONCLUSION

The study of visitors' time-space activity in the various destinations presented in this chapter added to our knowledge of spatial and temporal behavior of visitors. This knowledge can and should be exploited to better regulate tourist flows throughout the destinations studied.

Thanks to the GPS- or cellular-derived information, tourists may now be encouraged to visit previously deserted parts of a destination, from the scale of a miniature park to the national scale. Tourists can be prompted to visit popular attractions at specific times in order to reduce congestion and to allow them to fully benefit from their time in the city. Such information can also facilitate decisions as to where to set up new attractions and where to promote private-sector tourist services. In each and every case, the result of research will be the reduction of congestion in hitherto

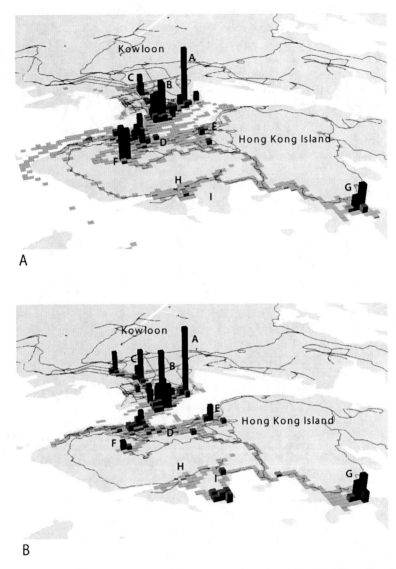

Figure 8.14 Aggregative analysis of first-time and repeat visitors in Hong Kong.

overcrowded, over-exploited areas and the general enhancement of the physical and social carrying capacity.

Lew and McKercher (2004; 2006) have suggested that urban tourist flows clearly have a tendency to spread themselves unevenly, both spatially and temporally. As a consequence, while the more popular sites and access routes in a destination often suffer from over-crowding and severe congestion, others are severely under-exploited, a state of affairs that points to a grossly inefficient use of economic and social resources and one which is ultimately unsustainable as well. There is clearly an urgent need for tourist management schemes designed to maneuver visitors around destinations in a more rational way. Such schemes would doubtless benefit from tracking-technology-based studies that are a remarkably efficient means of collecting a mass of high-resolution data on the spatial and temporal behavior of tourists.

9 Ethical Questions and the Tracking of Tourists

> In the future our cell phones will tag and track us like FedEx packages, sometimes when we are not aware. (Levy 2004, 81)

The question of privacy in relation to the increasing use in daily life of such devices as cellular phones, devices that are actually, inherently, amongst other things tracking devices (see Chapter 4), has become more pronounced during the recent decade (Curry 2000; Fisher and Dobson 2003). This concern has been widely debated by the popular media, the academia, and courts and lawmakers. However, in most countries, legal systems have failed to tackle the question fully and, as such, the question is of considerable and global importance (Renenger 2002). In this chapter we will relate to the specific ethics and privacy issues that arise from using advanced tracking technologies as research tools. (For further reading regarding general ethics and privacy concerns in social science research see, for example, Mertens and Ginsberg 2008 and Israel and Hay 2006.)

Over the course of this book, we demonstrated the abilities and potential of advanced tracking technologies for implementation in tourism research. These methods can provide a richer understanding of the mobility patterns of tourists in time and space and their impact on destinations. However, the use of such technologies, which are able to obtain the exact locations of research participants at any given moment, can cause infringements on the privacy of the participants and adds a geographical dimension to the surveillance society (Lyon 2001) and the ability to better track the digital individual (Curry 1997). This raises different questions about ethical and moral issues arising when conducting research that involves such technologies. One of the most prominent among these issues is whether, and to what extent, such research projects impinge on their participants' right to privacy.

The need to undertake research in a manner that will protect the privacy, dignity, and well-being of its subjects is obviously a concern to anyone engaging in any kind of study. The need to receive the approval of an ethics committee for research that involves humans is clear, accepted, and part of any research design in the social sciences. In the specific case of studies of the spatial-temporal activities of tourists, however, we believe that the potential of privacy abuse from the point of view of the participant is less of a problem due to the fact that people being tracked are monitored for a limited and well-defined time period and are not in their regular habitat. In our experience in various tracking projects in tourism and other areas, tourists are less reluctant to participate in such surveys.

Due to the novelty of using tracking technologies in tourism research, it has not yet received any attention in academic literature. For example, Fennell's seminal book on Tourism Ethics (2006) did not tackle this issue. Therefore, we thought it essential to include in this book a chapter exploring in more detail the concept of privacy in general and giving recommendations on how to tackle this issue in tourism research involving tracking technologies.

PRIVACY

The concept of privacy as a value that civic institutions like the law should promote emerged only relatively recently. Most accounts trace the origins of the modern debate on privacy to an 1890 article entitled "The Right to Privacy," published in the Harvard Law Review by Samuel Warren and Louis Brandeis. Prosser (1960, 383) offers the standard version of the article's genesis:

> In the year 1890 Mrs. Samuel D. Warren, a young matron of Boston, which is a large city in Massachusetts, held at her home a series of social entertainments on an elaborate scale. . . . Socially Mrs. Warren was among the elite; and the newspapers of Boston, and in particular the Saturday Evening Gazette, which specialized in "blue blood" items, covered her parties in highly personal and embarrassing detail. . . . The matter came to a head when the newspapers had a field day on the occasion of the wedding of a daughter, and Mr. Warren became annoyed. It was an annoyance for which the press, the advertisers and the entertainment industry of America were to pay dearly over the next seventy years.

Deeply aggrieved, Samuel Warren enlisted his old law school classmate, and eventual Supreme Court associate, Louis Brandeis. Warren and Brandeis readily acknowledged that no legal doctrine existed to redress the harm suffered by the Warren clan. Thus, instead of bringing a suit, the two young attorneys fought fire with fire and published their article.

Piecing together concepts from diverse legal subjects—property, contract, defamation, and Constitutional law—Warren and Brandeis extrapolated the existence of a "right to be let alone" (Warren and Brandeis 1890, 195). The proposal was embraced eagerly by the masses and eventually gained recognition by the judiciary.

Several basic, but important, observations about the nature of the right can be drawn from a reading of the article. Firstly, the perception of privacy as a value is tied inextricably to technological development. "Recent inventions and business methods call attention to the next step which must be taken for the protection of the person. Instantaneous photographs and newspaper enterprise have invaded the sacred precincts of private and domestic life" (Warren and Brandeis 1890, 195). Privacy, then, represents a legal response to technologies that threaten to make virtually every aspect of people's lives observable and publicly known.

Secondly, the version of privacy urged by Warren and Brandeis seeks to preserve notions of civility and manners of a distinctly polite society (Post 1989). Warren and Brandeis's proposal was spurred by photographs of lavish society balls, not by Jacob Riis's contemporary and equally intrusive photos depicting New York slums, to use an extreme example. Moreover, Warren and Brandeis perceived the essence of this peril lying not within the technologies themselves or the press but in the fawning eyes of the lesser classes. The newspapers and cameras were merely filling a demand "[t]o satisfy a prurient taste" and "[t]o occupy the indolent" (Warren and Brandeis 1890, 195). Privacy, as originally conceived, operated as a privilege of the elite class to insulate it, as necessary, from the scrutiny of commoners.

Belying its historical specificity and its rarified origins, the right to privacy is often described in countries that respect the human rights of their citizens as a cornerstone of every individual's relationship with government and with others. Additionally, in the theater of international law, the right to privacy has been elevated to a fundamental human value via its inclusion in the Universal Declaration on Human Rights (article 13).

Of course, this universalization of the privilege could only have occurred if privacy proved to possess a much broader appeal than the incarnation described by Warren and Brandeis. The doctrine of privacy would not likely have survived, much less expanded into what it is today, had it only been a defense against high-society paparazzi. This leads inevitably to the question of why privacy is so valuable in contemporary society. Responses to this query can be classified in two groups: those who understand privacy as an independently useful good and those who consider privacy a secondary good that helps people to achieve other social goals.

It is not difficult to incline toward the former view. Most people can sympathize with the shock or disgust experienced upon learning that one's privacy has been violated, for example, by the unwanted disclosure of personal facts. The intensity of this feeling suggests that it is indeed unrelated to secondary considerations. However, this position offers less analytical interest: "We could regard [privacy rights] purely as consumption goods the way economic analysis regards turnips or beer . . . but this would bring the economic analysis to a grinding halt, because tastes are unanalyzable from an economic standpoint" (Posner 1978, 393–394).

Turning to the second theory, a number of secondary benefits have been proposed to explain the value of privacy. The following discussion will list just a few of these and consider their applications to the specific problem of tourist tracking.

One such theory of privacy is espoused by Merton (1968), who claims that privacy is necessary to mitigate the pressure of living in a society so thoroughly governed by rules. Without the protection of privacy, "the pressure to live up to the details of all (and often conflicting) social norms would become literally unbearable" (Merton 1968, 429). Privacy, according to this theory, affords a permissive space where minor or inconsequential transgressions of social and legal tenets may occur, while concomitantly the public integrity of those systems is upheld. This conceptualization of privacy gains even greater resonance when considered in the context of tourism. The functions of privacy and of tourism largely overlap. Many people travel to break up the regimentation of their daily lives, to escape the rules that dictate their professional and social relations, and sometimes even to sample experiences that would not be legally or socially permissible at home. If a certain level of privacy is not afforded to the tourist, these motivations to travel are acutely undermined.

Another explanation is that privacy, and more specifically the distinction between private and public spaces, helps order social systems and sets out a space in which a person can realize and express his or her identity. This theory has gone through several different, complementary treatments. Rachels (1975) explains that privacy is important "because our ability to control who has access to us, and who knows what about us, allows us to maintain the variety of relationships with other people that we want to have" (Rachels 1975). The ability to selectively disclose and withhold information about one's life plays a crucial role in defining and ordering personal relationships. Put plainly, a person will discuss very personal subjects with close friends but not with casual acquaintances; these decisions help people situate themselves within meaningful social networks.

Curry (1997) suggests a similar account of the advantages of privacy. "People, after all, become individuals in the public realm just by selectively making public certain things about themselves. . . . [P]eople adjust their public identities in ways that they believe best, and they develop those identities in more private settings" (Curry 1997). Selective disclosure not only permits a person to structure his social relations effectively but also aids him or her in the formation of an individual identity. The trajectory of Curry's argument makes sense, considering the dynamic between the company one keeps and identity.

According to this set of theories, too, a nexus exists between the functions of tourism and the benefits afforded by privacy. Tourists often choose to travel to enrich or cultivate their identities. Because privacy protects the conditions that encourage identity formation, the erosion of privacy in a tourism context would undermine one of the key incentives to travel.

Finally, we must consider the costs incurred by the adoption of privacy. Too often, privacy is discussed as a "cannot-lose" proposition, opposed only by faceless corporate and governmental entities bent on quashing the rights of individuals. In fact, privacy has real and substantial costs that are spread across the spectrum of society. The most prominent of the elements sacrificed in the name of privacy is the free flow of information. Privacy prevents us from learning certain information about others and from sharing information we have learned. Sometimes the disclosure of such information is meaningless, or even harmful, but often it can actually be helpful.

Different Types of Legal Privacy

Because the concept of privacy has emerged and evolved largely within the context of ordered legal debate, and relatively recently, one might suppose that the contours of this particular right have been fixed and easily defined. However, the opposite has long been the case. Political philosophers have struggled to formulate a definition of privacy that both conforms to societal expectations and accommodates the interests at stake for years (Thomson 1975).

In the United States, the legal debate encompassing privacy has taken a similarly disjunctive course. As such, the concept of privacy consists not of a single right that can adapt to diverse fact patterns. Rather, privacy has developed into a series of distinct protectable interests. For present purposes, legal privacy can be divided into three basic categories: (1) fourth-amendment privacy; (2) common-law privacy; and (3) statutory and regulatory privacy. The doctrines that govern these rights vary substantially, with little to unify them conceptually beyond the label "privacy." Each protects different interests against different types of actions taken by different sorts of defendants. Moreover, each seems to borrow from or allude to concepts from pre-existing legal regimes. For example, many of the doctrinal elements of common-law privacy appear to derive from the ancient tort of defamation. These ubiquitous references to older, better established areas of law illustrate at least one characteristic common to all legal approaches to privacy: The struggle to address the problems of rapidly changing and complicated technologies and institutional conduct by employing an aging set of legal guidelines.

PRIVACY AND ETHICS ISSUES IN RESEARCH USING TRACKING TECHNOLOGIES

Unsurprisingly, then, little within the current law pertains directly to the use of land-based or satellite tracking systems. Nonetheless, one particular statute (and its implementing regulations) merits discussion for its applicability to tracking technologies: the Wireless Communications and Public

Safety Act (the WCPSA). Although the WCPSA is not overtly concerned with privacy issues, it has nonetheless shaped the way the government limits the sharing of GPS data. The WCPSA provided the Federal Communications Commission with congressional encouragement to continue to develop and implement the then nascent "enhanced 911" (E911) infrastructure.

The enhanced 911 scheme mandated that wireless providers develop their network capabilities to pinpoint the location of 911 calls made from mobile phones so that they may be relayed to emergency services. As we saw in the first chapter, the FCC has set a timetable for wireless carriers to implement the capabilities and establish accuracy requirements depending on the type of technology used (device-based or network-based; Federal Communications Commission 2006).

Even more pertinently, the WCSPA expanded the definition of "customer proprietary network information" under the Telecommunications Act to include information pertaining to the location of mobile network use. The Telecommunications Act places restrictions on how customers' proprietary network information can be shared. Telecommunication carriers are only permitted to disclose such information either with the consent of the customer or as related to its provision of services. Therefore in most cases mobile providers may not distribute locational information about their consumers.

This protective scheme does contain a few notable loopholes. Firstly, the restriction on disclosure is limited to "individually identifiable" information, so it does not apply to information that has been anonymized or where the personal identity of the customer is not discernable. Secondly, mobile carriers have been afforded leeway to obtain customer consent through fingerprint service agreements.

As already noted at the beginning of the chapter, in our view conducting tourism research with tracking technologies is less problematic in terms of potential privacy abuse for participants, because they are monitored only for a limited and well-defined time period in a destination that is not their regular habitat. In our experience, tourists were less reluctant to participate in studies than were other types of subjects. However, the privacy issue is not the only ethics issue in research and, in any case, the obligation of the researcher is to conform to certain moral and ethical principles.

Privacy and ethics issues differ in two types of research using tracking technologies in terms of the participants: (1) Direct Participation, in which participants are asked individually to take part in the study and give their consent. As will be explained later, there is an additional subdivision within this category between cases in which designated tracking devices are given to participants and cases in which subjects are tracked using their own tracking devices (i.e., their cellular phones). (2) Indirect Participation, when participants are tracked through the operator of a cellular provider, as was done in several cases described in the previous chapter.

Direct Participation

Two main avenues exist for drafting a participant to a tourist-tracking study in terms of the tracking device used. In the first, the researcher gives a device to the participant and it is returned at the end of the tracking period; in the second, the researcher relies upon a tracking device (usually a cellular phone) already being used by the potential participant. This second method might sound somewhat unusual, but it was practiced successfully by Ratti et al. (2005) in the city of Graz in Austria: visitors to an exhibition were asked to send a certain SMS message to the cellular operator, which resulted in their being tracked in the city. Thus researchers were able to see the aggregate pattern of all visitors who participated in this experiment in the exhibition itself.

When using either method, one common concern arises: What information should the participants receive before they consent to participate in the study? In our view three issues should be spelled out clearly: (1) the purpose of the study; (2) the fact that the identity of the participant will be concealed and that the information collected will not be transferred to a third party without his or her consent; and (3) instructions on how to abort the function of the tracking device at any given moment should the participant wish to stop participating in the study for any reason.

One aspect that is relevant only in cases in which the researcher gives a device to the participant is the need to assure the subject that the device will not harm his or her health in any way. For example, before giving out a device it is the responsibility of the researcher to investigate whether the device can potentially harm people with pacemakers or pregnant women. It is obvious that devices conforming to the highest health standards should be used, but it would be wise for the researcher to be very cautious in the choice of device and type of participants excluded from a certain study due to even the slightest danger to their health.

We would therefore advise that technical, legal, and medical issues be taken into consideration when designing research involving tracking devices. This is also true of other types of research involving the distribution of any kind of devices.

Indirect Participation

In Chapter 8 we presented the possibilities of mass tracking of cellular phones using different methods. One of them was passive positioning. This method does not involve problems of privacy because specific cellular phones are not tracked; rather a statistical analysis of activity on different network transceivers is conducted. The other method of analysis of cellular tracking is detecting the location and migration of a group of devices over a given period of time on the networks' different transceivers. Using this method, there are several cases in which a privacy breach is possible.

The first potential pitfall occurs in cases in which a relatively low number of people are being tracked. In such cases, individual participants might be identified. The second is the possible identification of the participant through his or her cellular phone number. For example, when a participant is the only person located within a cell, his or her identity can easily be uncovered. In order to protect the privacy of cellular network subscribers when data that have been created by them are used for research, it is essential to maintain a degree of uncertainty regarding the identity of the subscriber. There remains a tension between the researcher's desire to collect as many details as possible regarding the research subjects and his or her commitment to guard the participants' privacy. Each new detail introduced to the research describing a participant increases the chance that an observer might recognize the person within the full data set. Uncertainty is therefore essential when dealing with spatial data that are connected to people and can impede on their privacy.

There are two types of sources for uncertainty (see Figure 9.1, taken from Duckham et al. 2006): intentional uncertainty, created in order to make the identification of an individual more difficult than it would be otherwise, and uncertainty created unintentionally as a result of the inaccuracy of a method or technology.

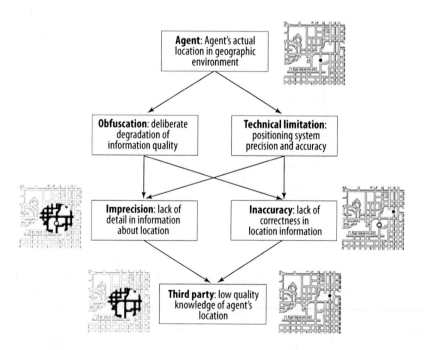

Figure 9.1 Summary of location privacy scenario. Source: Duckham et al. 2006.

Reades (2008) suggested using finite state machines as a concept on which a tool that masks privacy can be built. When the data collected from the cellular network enter a "state" that is viewed as dangerous and can cause a breach of privacy, actions are taken throughout the whole data set to change the situation and return to a state in which there is no danger of a breach of privacy.

Two steps must be taken in our case: First, researchers must verify that the data that the phone company gives away do not contain the numbers of the devices. This is something the phone company should and probably will do in any case, since it is very careful not to be liable in any privacy breach regarding its customers; in fact, this is why such private companies may be reluctant to provide such data in the first place. This is generally true, though we did see in the previous chapter that researchers have managed to get access to this kind of data and that cellular companies offer to sell data about aggregative patterns of their customers, mainly for transportation management companies. The second step that must be taken is to check that a sufficient number of tracks are present in each geographic cell so that identification of a specific individual is not possible.

Privacy issues, as we saw, have evolved over time, leading to a state in which the protection of the privacy of the individual is an uppermost value in most Western countries. This must be taken into account when designing studies which examine human behavior. In the field of tourism research, the issue of maintaining the privacy of the participants is important and an essential part of any study. However, we are confident that, by following strict guidelines and avoiding the pitfalls outlined in this chapter, researchers can and should use tracking technologies of all kinds to learn more about the time-space activities of tourists.

Part IV
Concluding Thoughts

10 Conclusion

A Trillion Points of Data: How tracking cell-phone users via GPS could do for the real world what Google did for the virtual world. (Sheridan 2009, 32)

This book documents an early stage of the application of tracking technologies in tourism, which is one of the first fields in which these methods have been implemented, probably because of the centrality of mobility in tourism. We hope that we managed to infuse the reader with some of our enthusiasm on the subject. If properly applied in tourism research, tracking technologies could be revolutionary, pushing the boundaries of tourism studies and improving policy making, planning, and tourism management. While touching upon the challenges and problems involved, this book presents these technologies' potential as tools for gathering data on the spatial whereabouts of tourists at any given time, potential that must still be developed by further empirical research. Concluding this book is a difficult task. It seems that every day brings with it advances in this new and emerging field; putting down the pen is thus difficult to do.

The book offered a systematic review of the field of tourist time-space activities and the traditional methods of measuring and visualizing them. It followed this overview with a presentation of the central tracking technologies available. The dilemmas and insights regarding research design followed and then the two primary angles for analysis of the high-resolution data obtained by these tracking technologies: focusing on the tourist and focusing on the destination. The previous chapter dealt briefly with the ethical issues that are at stake when using these technologies and the appendix will highlight some subjects related to the analysis of data using geographic information systems (GIS).

The development of GPS receivers and mobile phones is a very dynamic domain; in addition, these devices have recently become easily accessible and more affordable. Though research employing these methods is not yet mature, in our view enough has been accomplished at this stage to be documented. The possibilities for implementation of these technologies for the purposes of research are increasing and have the power to change tourism mobility research as it stands today. We hope that this book will serve as a helpful guide for researchers who wish to implement tracking technologies in order to obtain objective and high-resolution data about the time-space activities of tourists.

Providing extremely accurate data in time and space, these technologies are opening up new, previously unfeasible, lines of inquiry. In practical terms, the high-resolution data thus obtained could be used, among other things, to monitor and regulate the carrying capacity of tourist sites more rationally, improve the allocation of accommodation services, reduce friction between tourists and local populations, and assist in planning tourism transport infrastructures. All in all, an important point to be clarified is that these technologies do not replace questionnaires, diaries, or interviews, which will remain an important source of information on the activity—and especially the motives underlying the activity—of tourists. Instead, the new technologies will complement, add to, and enrich the findings of these more traditional research tools.

The examples presented in this book were mostly carried out within a limited scale both in terms of the numbers of participants and the time that they were sampled for. In the future, we will no doubt see research projects that are executed on larger scales covering wider geographic regions, including bigger samples over longer periods of time. Cities and countries that flourish on tourism will be able to gain knowledge about the flow of tourists throughout their vicinities by carrying out ongoing surveys of the locations of tourists within their boundaries. Visitor satisfaction surveys and questionnaires about tourism spending patterns will be considered incomplete without information on where the visitor has spent his or her day and spatial patterns of satisfied visitors will become an industry standard. Tourism planners will be able to consider the impacts of different development scenarios more accurately and objectively. This will enable a much higher degree of efficiency in the management of tourism destinations of different scales.

Over the course of the book we presented two very different sources for obtaining data regarding spatial activity: data derived from cellular networks and data collected using personal tracking devices carried by participants in a study. We presented the advantages of each technology and tried to guide the reader in understanding the aspects that must be considered before deciding on a technology to be employed in a specific type of research. Experience will teach us whether both approaches to data collection prove practical for use in large research environments. Which collection technology will prove to be both scalable, allowing a researcher to take on big research projects, and broad, supplying the researcher with a richness of information? At this point, the ability of additional researchers to make the necessary connections with cellular communications providers and obtain high quality data from them is still unknown. Only more research in the field can give us answers to these questions.

FUTURE AGENDAS FOR RESEARCH

One open question that was not dealt with in this book and still must be investigated is: Do visitors, once they know they are being followed, change their activity, and, if so, how? This question should be further explored in empirical studies. However, even the more traditional methods for gathering information about the time-space activities of tourists raise concerns as to whether reports are inaccurate (in questionnaires and time budgets) and whether tourists change their patterns of activity due to participation in a study, feeling that they must behave in a certain way.

An additional direction of inquiry that has yet to be explored is the combination of tracking data with additional digital sources of data. Items that can be added to a tracking kit and carried by the participant can include, for example, sensors that measure excitement and physical effort. Sensors can also be placed throughout a destination to measure, for instance, noise levels, pollution, and other qualities of the environment.

Another remaining question relates to the large amounts of data that are accumulated when using these methods, which will require the development of more algorithms to enable automatic scripts to analyze the data in a fast and practical way. It would be beneficial if at some point the software being developed today by different research teams around the globe on an ad hoc basis could be standardized so that common measures are developed. More importantly, this could lead to an increase in the number of researchers in the field, because the current challenge of analyzing the data no doubt limits the number of prospective users of these methods.

The large amounts of accurate data collected using these methods will make possible the development of new theories regarding the spatial activity of tourists, new understandings of tourists and their influence on destinations, and more accurate ways to calculate both physical and social carrying capacity in order to ensure the sustainability of tourism within a destination. These developments were not possible using the data sources that were available up until the introduction of these technologies.

Further research is necessary in order to develop an understanding of what patterns of spatial activity are beneficial both to the destination and to tourists visiting the destination. Recognizing patterns of spatial activity throughout a destination that are not beneficial, that may lead to damage of delicate attractions (such as historical locations) or infrastructure, and that may cause frustration and displeasure to the tourists can be of great assistance in the smooth and efficient management of tourist destinations. Being able to read into the spatial activity, to understand the different meanings that these patterns of behavior have for tourist destinations, and to project the possible results of various interventions are essential skills and can be developed through research. We have the ability to recognize areas of high and low activity and even to classify destinations into different types of space according to spatial activity, but we do not yet have the knowledge to identify over-usage, when activity levels have reached a point that is no longer beneficial for the destination.

The potential improvements in the implementation of tracking technologies for tourism research and practice mentioned above could enhance the benefits of using those methods for researchers and practitioners alike. The rapid advances in tracking technologies and the growing possibilities in implementing them for use in different areas in general and in the realm of tourism research in particular, as were elaborated in this book, leaves no doubt that the future of tracking technologies in tourism will be exciting and dynamic, providing researchers with invaluable insight and information.

Appendix
Integrating Data Obtained from Tracking Devices into Geographic Information Systems

This appendix explains the process of integrating GPS data into a Geographic Information System (GIS) so that the data can be analyzed spatially. Outlining the principles of integrating GPS data into GIS systems, the appendix gives practical tips as to practices that can shorten the process and lead to better results. The examples cited in this section use GIS software manufactured by ESRI® (widely used GIS software). Explanations of concepts and key challenges that must be overcome when importing GPS data into GIS software are provided and can be applied to other GIS software available on the market.

The basic data that are collected using GPS (or any other tracking device) as a tool for collecting data in all research is a set of points. Each point has three attributes: the X coordinate, the Y coordinate, and the time. The X and Y coordinates allow us to position the point on a map; the time of the point is needed in order to determine what amount of time the point represents and what the sequence of events was. Different devices have distinct settings and use various processes of saving collected data. Some devices have the option of saving a point for a set unit of distance traveled (for one example, see the Magellan GPS receiver used in experiment number 1 in Chapter 6) and others save a location at a temporal resolution that can be determined by the user. Regardless of these differences between devices, the ability to manipulate the collected data in order to analyze them is essential for conducting any research based on the data gathered.

GPS Protocols

Different manufacturers allow saving the data collected using GPS receivers in different protocols or formats. Some of the options are formats that have been developed by the manufacturers and allow the data to later be viewed or played back using tools provided by the manufacturer. Other options are protocols that are recognized internationally as the industry's standard.

The primary protocol used on virtually all GPS receivers is the NMEA 0183 format. This is a format defined by the National Marine Electronics

Association (NMEA). When using the NMEA format, a plain ASCII comma-delimited text file is saved. The first field in each NMEA sentence, called a data type, contains a code that defines the way the rest of the sentence should be interpreted. Each data type defines the meaning of the data found in the different fields of the sentence.

Another data-saving option involves saving specific information requested by the researcher in a text file. Many receivers have a user interface that allows one to choose what parameters will be saved and in what temporal resolution. The parameters that can be saved are the parameters included in the NMEA format. Using this option, the receiver extracts the specified values from the NMEA data and saves it as a text file (mostly in comma-separated values [CSV] format). The original NMEA data will not be saved.

Understanding NMEA

$GPGGA,152926,6027.8259,N, 2225.6713,E,8,09,2.0,44.7,M,20.6,M,,*79

Above is an example of an NMEA sentence containing time and location. This is the most basic NMEA sentence and will always appear in an NMEA GPS file. The data type of this sentence is "GPGGA," as it appears in the first field. This data type is the fundamental and most important data type, necessary in order to extract the three needed parameters of longitude, latitude, and time; unlike other data types that are not always implemented, this data type is implemented by all receivers and will always be available to be saved.

The second field contains the time in universal time coordinated (UTC) format. This format of time contains six digits: two for the hour, two for the minute, and two for the second. Parts of a second can be included using a decimal point after the two digits used for the seconds. This time represents the time in Greenwich which lies on the zero longitudinal meridian, also referred to as GMT (Greenwich Mean Time).

The third field contains the longitude of the position, followed, in the subsequent field, by an "N" or "S," which indicates within which side of the globe the longitude is located. The fifth field contains the latitude followed by an "E" or "W," indicating which side of the globe the latitude is located within.

The remaining fields include information regarding the quality of the GPS reception and fix as well as parameters such as whether or not there was a fix on the satellites and how many satellites were used to obtain the location. The ninth field in the sentence contains the altitude of the point.

Both latitude and longitude contain four spaces to the left of the decimal point and up to four places after the decimal point. These coordinates are in degrees, minutes, and decimal minutes (DM.m) and need to be converted into degrees and decimal degrees (D.d) so that they can be imported into the GIS. The formula for this conversion is: Divide the minutes and the decimal minutes by 60 and add the result to the degrees.

For example:

DM.m → D.d (3214.5462 → 32.2424)
Divide M.m by 60 to get .d
(14.5462/60 = 0.2424)
Add .d to D to get D.d
(32 + 0.2424 = 32.2424)

Importing Data to GIS

Theoretically speaking, one should be able to import a comma-separated text (CSV) file directly into a GIS and plot the points using a "display x,y" command. In practice, when importing many points (and data collected using GPS can grow into large data sets very quickly), CSV files are difficult to import without errors. One way to attempt to simplify the process is by inserting the CSV file containing the data into a database (Microsoft Access, for example) and then importing the database into the GIS platform. The advantage of importing the data this way is that creating the database prior to importing the data into CSV allows for better control of the data being imported. When creating the database one also has better control over the attributes of the different fields (nominal, integer, real number, numeric precision, etc.).

Accuracy and Precision

These terms have meanings that are similar but not identical. Understanding the different meanings will help us discuss errors, their sources, and how to address treating them.

Precision is defined as how close measured values are to each other. Accuracy, on the other hand, describes how close the measured values are to an actual goal. The best way to understand this is to visualize throwing darts at a target. If the darts hit the target close to each other but away from the center, then we would say that the darts were thrown with high precision but low accuracy. If the darts hit the target in the center, we would say that the person who threw them was both precise and accurate. Darts that are scattered around the target are neither accurate nor precise.

In terms of GPS, precise measurements are obtained when a receiver is stationary and the points measured are close to one another. The precision of points obtained from GPS receivers is influenced both by the characteristics of the receiver and by characteristics of the GPS such as atmospheric conditions and satellite geometry.

Accuracy of measurements can be measured when a receiver is placed in a known location or moves over a known path. After plotting points on a map, the distance between the point on the map where the points should be and the point in which they are actually found is the accuracy. Accuracy is influenced by the precision of the GPS receiver, the accuracy of the mapping, and the way in which the points were plotted.

GIS Layers

In order to plot data that are collected using GPS devices, one must obtain base layers that can be used in a GIS system. GIS data are usually divided into thematic layers, each layer including one type of entity. The layout of the roads is the most basic information; roads and pathways are the main arteries people move through and are essential in understanding their spatial activity. Base layers must therefore include the layout of roads and possibly pathways. Other useful information includes attractions, buildings, parks, parking lots, and public transportation routes.

Three main sources for geospatial data exist:

- Digital data from government agencies;
- Digital data from private vendors; and
- Scanned maps.

Government agencies like census bureaus and mapping agencies create spatial-temporal data for their own uses and often offer the data to the public for a fee or as a free service. An example of GIS data that have been made accessible to the public is the GIS data made public by the U.S. Census Bureau. The Census Bureau's GIS layers, covering the whole United States, are known as TIGER (Topologically Integrated Geographic Encoding and Referencing system) and can be down loaded for free or for a small fee from the agency's website (www.census.gov/geo/www/tiger/index.html). The data found in these layers can vary greatly in quality and can be anywhere from exceptionally accurate and exact to sloppy and inaccurate. Not all countries offer GIS layers to the public as a service; in some countries this resource is not available to researchers.

Many private companies offer GIS information for a fee. Some are global companies that offer information about a number of countries while others are local companies whose expertise is in the mapping of a specific country or area. Companies offering location-based services and GPS navigation systems are the main clients of these companies; they integrate the GIS data they buy into the systems that they build. Navteq (www.navteq.com) is a company that offers GIS layers at different resolutions and in many different locations with world-wide coverage; their layers are widely used by many organizations.

The last source of geospatial data, scanned maps, is the least recommended of the three but is important to keep in mind in cases in which no other sources of data can be obtained. Scanned maps are often less accurate than digital spatial information due to the cartographic limitations involved in printing a map. The scale of the map, along with the purpose for which it was made, are the main criteria that determine its accuracy. After the maps are physically scanned they must be aligned correctly within the GIS. Once this is done they can theoretically be used at any scale—but they will not be accurate when used on a scale smaller than the original scale for which the map was produced. One example of inaccuracies are the width of roads, which are often much wider than the actual road (on a 1:50,000 map, small urban roads can often be represented up to two times their actual width).

Coordination Systems

The GPS system obtains locations using the latest revision of the World Geodetic System (WGS) from 1984. WGS is a world-wide coordinate system that uses degrees, minutes, seconds, and fractions of seconds to specify a point on the surface of earth. The system is used in GPS satellites and receivers because of its high accuracy and its coverage of the whole globe. Being a global system, its coordinates are not projected onto a flat surface; they remain on the round shape of the ellipsoid that represents earth.

The main drawback of a coordinate system that is not projected is the lack of ability to accurately measure distances on the surface. The distance that the degrees, minutes, and seconds represent is different in different locations throughout the system. Close to the poles the degrees are closer together and at the equator they are the most spread out. In order to spatially analyze the collected data, the points collected need to be projected into a local coordinate system.

GIS software is capable of projecting geographic data. Generally, it will have a number of different possible options for projecting the data in addition to the default program used when no other method is specified.

Table A.1 Comparison of Different Base Layers

	Coverage	Accuracy	Spatial resolution
Digital data from government agencies	Local (by country)	+	+
Digital data from private vendors	World-wide	+	+
Scanned maps	World-wide, pending physical availability	-	-

Knowing what projection method is used by the software and being able to control the method used are crucial when aiming for the highest possible accuracy. When discrepancies between points collected and the map layers are found, going back and examining the way that the data was projected is essential. This can eliminate the source of many errors.

Many GPS receivers have a built-in ability to record coordinates obtained in a local coordinate system and not in the WGS '84 system which is used by the GPS. This option should be treated with care and possibly avoided altogether. When the receiver records coordinates in a coordinate system other than system originally used by the GPS system, the receiver is actually projecting the points from the WGS coordinate system on a local coordinate system. This projection is done using the default projection programmed into the receiver by the manufacturer, leading to two main drawbacks: The first and most troubling is the loss of control that occurs after this conversion has been done. If the points do not stand on the map properly, no means exists to backtrack and change the conversion parameters in order to achieve higher accuracy. The second drawback is the lack of understanding that the researcher has as to the process that the points underwent. Adding data from an additional source at a later time will be impossible, as will be improving the accuracy in any way.

Treating Misplaced and Missing Points

Once the data that have been collected have been successfully integrated into a GIS it is time to perform quality control. The first step is to inspect the data and verify that they are laid properly on the base layers of streets and buildings. Streets have distinct shapes that should be compared to the points to assure a good fit. If the collected points do not fit the base layers, the researcher must backtrack to verify that the projection was done in a reliable way.

Another possible source for layers not fitting can be found within the base layers themselves. Who created the base layers? What is the accuracy of the layers? What coordinate system were they created in? Were they projected? How and by whom? All of this information is needed in order to pinpoint the possible source of inaccuracy and, with luck, rectify the problem.

After the researcher is confident that the GPS data fit the base layers properly, he or she can scan the data for misplaced or missing points. Problems with the collected data can be in temporal or spatial dimensions or in both dimensions simultaneously: Spatially problematic points are points that are misplaced; temporal problems in the form of periods of time when locations were not obtained by the GPS receiver. Locations may not be obtained for several reasons, the most obvious being because the receiver was shut down.

Locating misplaced points is easiest to do by looking at one track at a time and playing the track using a temporal tool like the "Tracking

Analysis" or basic temporal tools available for use with the most recent versions of ArcMap (ArcMap 9.1 and above). Filtering the data by selecting only the points that were obtained when there was a fix on the satellites can help in eliminating some of the points that are not placed correctly. Misplaced points are easily recognizable when the data are displayed temporally; the misplaced points briefly "jump" out of the location of the rest of the points and then return to where the points are really located. This is done at a speed that is too fast for a human to move (when not riding in a vehicle) and does not show movement along the roads and paths like a person would move.

Another strategy for verifying the projection is to understand the locations that are more prone than others to having misplaced points and to examine these locations especially carefully. Locations that are prone to having misplaced points may be indoors, with problematic reception of signals emitted by GPS satellites. Many times, the GPS receiver within these locations will record points that have been misplaced instead of simply recording no points at all. Understanding the behavior of the GPS receiver in locations that do not have GPS reception can make the job of locating misplaced points easier.

Once misplaced points are pinpointed, the researcher must decide what to do with them. The two basic options are to delete the misplaced points or to move them back to their presumed locations. Unless one is able to come up with a sound method for relocating points in a way that ensures that they are relocated to their correct place, deleting misplaced points ensures that the integrity of the spatial database is not compromised. Data that are missing as a result of the receiver turning off or losing power might not be suitable for analysis based on the length of time that is missing. In the case of missing data, it is not possible to reconstruct the missing points without compromising the database. It is wise to set a temporal threshold, in which the whole track is marked as not suitable for analysis when more time is missing than the threshold allows.

Spatial Analysis

Once the data are successfully integrated into a GIS, the next step is analyzing the data. Generally speaking, two main types of analysis can be conducted. The first type of analysis examines the movement of the tracked person; the second type examines the allocation of time within space. Movement analysis of tourists' spatial behavior can include calculating the speed of movement, the mode of transportation, and the routes taken. The study of time allocation addresses how people allocate their time as tourists within the destination that they are visiting. Both types of analysis can be conducted at different levels of aggregation. The table below delineates the possibilities of analysis at different aggregation levels and explains the meanings of the combination between analysis and level of aggregation.

Table A.2 Analysis at Different Aggregation Levels

	Movement	*Time Allocation*
Individual tourist (one track)	Visualizes the track, can help in "jogging" the memory of participants as to what they did throughout the research period.	Can serve as a tool for an in-depth interview.
Group of tourists, grouped using a common criterion such as age, nationality, etc.	Identifies routes that are preferred by people in the analyzed group; identifies areas that people wander through vs. areas of functional movement.	Identifies patterns in time-spending within the group and between different groups.
All collected tracks	Identifies routes that are used by tourists; enables analyzing the carrying capacity of these routes (for example, small alleys and walkways in historic cities).	Analyzes how all of the tourists allocate their time.

Representing the Destination

The most popular way to represent space in policy-making fields has always been by drawing polygons that represent zones (Thurstain-Goodwin 2003). Zones can be determined by the data collection agency (as in the case of data that are collected by a central census bureau), determined by the local authority (as in the case of a park or another closely managed destination), or determined by the researcher him- or herself.

Using zones to analyze data assumes that within the zone, the space is uniform in some respect. A good example of uniform space is the division into polygons of the attractions in an amusement park (for examples see Figure 8.5 in Chapter 8). This division, created by the park management, was in line with the research goals and helped serve the study's purposes. The division of Akko into polygons as presented in Chapter 7 was done as part of the research project, recognizing areas within the city that have the same functionality. When the division is done this way there can be more than one possible division. The division process should be a premeditated and careful one.

Calculating Allocated Time

Once the location of the study is divided into polygons, the collected points that describe spatial activity must be joined to the polygons. This is most easily achieved by joining the two layers based on the location of the points.

After the layers are spatially joined the points in each polygon must be summed. The most obvious value to sum would be the duration of time spent within the polygon and this can be done quite easily. Other numeric values can be summed or averaged—such as average walking speed within the polygon or the number of visitors that have passed through the polygon—resulting in maps with different meanings.

Another approach to representing the destination is analyzing the space of the destination by spreading a net of equally sized cells over the destination and examining the activity within each of the cells. Equal division of space can be achieved by using either a vector data structure or a raster data structure. When applying a vector data structure, square polygons of equal size are created; these polygons are referred to as a fishnet. A raster data structure is more efficient computation-wise and is designed to manipulate large amounts of cells over large areas, though this must be done within the strict definitions of this data structure. Fishnets are more flexible but can pose computation difficulties when a large number of cells are created within the net.

Analyzing Movement

The points that have been collected using GPS include both stationary points and points sampled while the participant was moving. The points sampled when the person was not moving outdoors—for example, when he or she was sitting at a café or inside a store—are usually obtained in a certain radius of the point where he or she is actually located (for more information on GPS accuracy, see Shoval and Isaacson 2006).

Calculating accurate information regarding the movement of the participants requires us to filter the points that do not describe movement. Filtering these points allows us to achieve two main goals. The first is the attainment of a general understanding of the participants' basic spatial behavior. How much time did the person spend at home? How much time did he or she spend at other destinations, such as work or a social club?

The second goal is the calculation of parameters that describe the participant's movement. How fast did he or she walk? What distance does the participant walk every day? When does the participant choose to drive in a car or use public transportation? Not filtering these points will result in including as movement the shifting points that are recorded when the person is stationary. A person who stays home all day will seem to be very active walking around the area in which he or she is sitting.

One way to analyze the participant's movement is by filtering the stationary points and classifying all of the points collected as belonging to either "nodes" or "tracks." This distinction is central and has molded the way the data are treated in these projects; many of the strengths and limitations result from it. Nodes are points that describe a time when the participant was stationary. Tracks are points that describe movement. When examining points classified as tracks, it is possible to calculate the speed of movement and possibly the mode of transportation. In the data set, each track is defined by having two nodes, one at either end. The transition between nodes is by movement through a track.

This appendix was not written as a guide for GIS analysis but, rather, was aimed at giving the researcher who is not familiar with GIS a basic understanding of the various stages involved in analysis of tracked data. This appendix gives the reader an understanding of the overall process, what needs to be achieved, and what challenges he or she may face in the process. If the researcher does not have a working knowledge of GIS, he or she will at least be able to oversee such an analysis by a GIS technician. The process explained here can be very time-consuming if done manually and can benefit greatly by automated processing. Such automated processing does not exist commercially and would need to be developed.

For further reading on integrating GPS data into GIS the following sources are recommended:

Logsdon, T. (1995) Understanding the Navstar: GPS, GIS and IVHS. New York: Springer.

Thurston, J., Poiker, T. K. and Moore, J. P. (2003) Integrated Geospatial Technologies: A Guide to GPS, GIS, and Data Logging. Chicester: John Wiley & Sons.

Taylor, G. and Blewitt, G. (2006) Intelligent Positioning: GIS-GPS Unification. Chicester: John Wiley & Sons.

References

Abbott, A. (1995) 'Sequence analysis: New methods for old ideas', *Annual Review of Sociology*, 21: 93–113.

Abbott, A. and Forrest, J. (1986) 'Optimal matching methods for historical sequences', Journal of Interdisciplinary History, 16(3): 471–94.

Abbott, A. and Hrycak, A. (1990) 'Measuring resemblance in sequence data: An optimal matching analysis of musicians' careers', American Journal of Sociology, 96(1): 144–85.

Abbott, A. and Tsay, A. (2000) 'Sequence analysis and optimal matching methods in sociology: Review and prospect', Sociological Methods & Research, 29(1): 3–33.

Adler, S. and Brenner, J. (1992) 'Gender and space: Lesbians and gay men in the city', International Journal of Urban and Regional Research, 16:24–34.

Ahas, R., Aasa, A., Mark, U., Pae, T., and Kull, A. (2007) 'Seasonal tourism spaces in Estonia: Case study with mobile positioning data', Tourism Management 28(3): 898–910.

Ahas, R., Aasa, A., Roose, A., Mark, U., and Silm, S. (2008) 'Evaluating passive mobile positioning data for tourism surveys: An Estonian case study', Tourism Management 29(3): 469–86.

Ahas R. and Laineste J. (2006) 'Technical and methodological aspects of using mobile positioning data in geographical studies', in K. Pae, R. Ahas, and Ü. Mark (eds.) Joint Space. Open Source on Mobile Positioning and Urban Studies, Tallinn: Postium.

Ahas, R and Mark, Ü. (2005) 'Location based services—new challenges for planning and public administration?', Futures, 37: 547–61

American Express (1989) 'Unique four nation travel study reveals travellers' types', News Release. London, 25 September.

Anderson, J. (1971) 'Space-time budgets and activity studies in urban geography and planning', Environment and Planning, 3(4): 353–68.

Appleyard S.F., Linford, R.S., Yarwood, P.J., and Grant, G.A.A. (1988) Marine Electronic Navigation, London and New York: Routledge.

Arbel, A. and Pizam, A. (1977) 'Some determinants of urban hotel location: The tourists' inclinations', Journal of Travel Research, 15(3): 18–22.

Arentze, T.A. and Timmermans, H.J.P. (2000) ALBATROSS: A Learning-Based Transportation Oriented Simulation System, The Hague: European Institute of Retailing and Services Studies.

Ashworth, G.J. and Tunbridge, J.E. (2000) The Tourist-Historic: Retrospect & Prospect of Managing the Heritage City, Amsterdam and New York: Pergamon.

Bargeman, B., Chang-Hyeon J., and Timmermans, H. (2002) 'Vacation behavior using a sequence alignment method', Annals of Tourism Research, 29(2): 320–37.

Batty, M. (2005) 'Agents, cells, and cities: New representational models for simulating multiscale urban dynamics', Environment and Planning A, 37(8): 1373–94.

Batty, M. Dodge, M., Doyle S., and Smith, A. (1998) 'Modelling virtual environments', in P.A. Longley, S.M. Brooks, R. McDonnell, and B. MacMillan (eds) Geocomputation: A Primer, New York: Wiley, 139–61.

BBC News (2007) 'Unanimous backing' for Galileo', BBC News, 30 November, http://news.bbc.co.uk/1/hi/sci/tech/7120041.stm.

Bell. D.J. (1991) 'Insignificant others: Lesbian and gay geographies', Area, 23: 323–29.

Bhat, C.R. (1996) 'A hazard-based duration model of shopping activity with non-parametric baseline specification and nonparametric control for unobserved heterogeneity', Transportation Research B, 30: 189–207.

Blair-Loy, M. (1999) 'Career patterns of executive women in finance: An optimal matching analysis', American Journal of Sociology, 104(5): 1346–97.

Bohte, W. and Maat, K. (2009) 'Deriving and validating trip purposes and travel modes for multi-day GPS-based travel surveys: A large-scale application in The Netherlands', Transportation Research Part C, Article in Press.

Borg, J. van der, Costa, P. and G. Gotti (1996) 'Tourism in European heritage cities', Annals of Tourism Research, vol. 23 (2): 306–321.

Bourdieu, P. and Darbel. A. (1991) The Love of Art: European Art Museums and Their Public, Cambridge: Polity Press.

Bowditch, N. (1995) The American Practical Navigator, Arcata, CA: Paradise Cay Publications.

Bowman, G. (1991) 'Christian ideology and the image of a holy land: The place of Jerusalem pilgrimage in the various Christianities', in J. Eade and M.J. Sallnow (eds) Contesting the Sacred: The Anthropology of Christian Pilgrimage, London and New York: Routledge.

Braun, B. M. (1992) 'The economic contribution of conventions: The case of Orlando, Florida', Journal of Travel Research, 30(3): 32–7.

Brown, B. and Chalmers, M. (2003) 'Tourism and mobile technology', Proceedings of the 8th European Conference on Computer Supported Cooperative Work (ECSCW 2003), Helsinki, 14–18 September, 2003: 335–54.

Burtenshaw, D., Bateman, M., and Ashworth, G.J. (1981) The City in West Europe, Chichester: John Wiley & Sons.

Burtenshaw, D., Bateman, M., and Ashworth, G.J. (1991) The European City: A Western Perspective, London: David Fulton Publishers.

Calabrese, F. and Ratti, C. (2006) 'Real time Rome', Networks and Communication Studies—NETCOM, 20(3–4): 247–58.

Canestrelli, E. and Costa, P. (1991) 'Tourist carrying capacity: A fuzzy approach', Annals of Tourism Research, 18(2): 295–311.

Carr, N. (1999) 'A study of gender differences: Young tourist behaviour in a UK coastal resort', Tourism Management, 20(2): 223–8.

Chadefaud, M. (1981) Lourdes, un Pélerinage, une Ville, Aix-en-Provence: Edisud.

Chapin, F.S. (1974) Human Activity Patterns in the City, New York: Wiley.

Cohen, E. (1972) 'Toward a sociology of international tourism', Social Research, 39(1): 164–82.

Cohen, E. (1979) 'A phenomenology of tourist experiences', Sociology, 13: 179–202.

Cohen, E. (1985) 'The tourist guide: The origins, structure and dynamics of a role', Annals of Tourism Research, 12(1): 5–29.

Connell, J. and Page, S.J. (2008) 'Exploring the spatial patterns of car-based tourist travel in Loch Lomond and Trossachs National Park, Scotland', Tourism Management, 29: 561–80.

Cooper, C.P. (1981) 'Spatial and temporal patterns of tourist behavior', Regional Studies, 15: 359–71.

Curry, M.R. (1997) 'The digital individual and the private realm', Annals of the Association of American Geographers, 87: 681–99.

Curry, M.R. (2000) 'The power to be silent: Testimony, identity, and the place of place', Historical Geography, 28: 13–24.

Dann, G.M.S. (1993) 'Limitations in the use of "nationality" and 'country of residence' variables", in D.G. Pearce and R.W. Butler (eds) Tourism Research: Critiques and Challenges, London and New York: Routledge, 88–112.

Dann, G.M.S. (1996) The Language of Tourism: A Sociolinguistic Perspective, Wallingford, UK: CAB International.

Debbage, K. (1991) 'Spatial behavior in a Bahamian resort', Annals of Tourism Research, 18: 251–68.

Dietvorst, A.G.J. (1994) 'Cultural tourism and time-space behavior', in G. Ashworth and P. Larkham (eds) Building a New Heritage: Tourism, Culture and Identity in the New Europe, London: Routledge, 69–89.

Dietvorst, A.G.J. (1995) 'Tourist behavior and the importance of time-space analysis', in G. Ashworth and A.G.J. Dietvorst (eds) Tourism and Spatial Transformations, Wallingford: CAB International, 163–81.

Djuknic, G. and Richton, R. (2001) 'Geolocation and assisted GPS', Computer, 34: 123–25.

Duckham, M., Kulik, L., and Birtley, A. (2006) 'A spatiotemporal model of strategies and counter-strategies for location privacy protection', Proceedings of the Fourth International Conference on Geographic Information Science, Schloss Münster, Germany.

Edensor, T. (1998) Tourists at the Taj: Performance and Meaning at a Symbolic Site, London and New York: Routledge.

Egan, D.J. and Nield, K. (2000) 'Towards a theory of intraurban hotel location', Urban Studies, 37(3): 611–21.

El-Rabbany, A. (2006) Introduction to GPS: The Global Positioning System, Norwood MA: Artech House.

Elgethun, K., Fenske, R.A., Yost, M.G., and Palcisko, G.J. (2003) 'Time-location analysis for exposure assessment studies of children using a novel global positioning system instrument', Environmental Health Perspectives, 111: 115–22.

Eurostat (2005) Europe in Figures: Eurostat Yearbook 2005, Luxemburg: Office for Official Publications of the European Communities.

Federal Aviation Administration (2007—updated) GNSS, Frequently Asked Questions–WAAS. Accessed 9 October 2007. http://www.faa.gov/about/office_org.

Federal Communications Commission. (On-Line) Enhanced 911 (retrieved, 6 June, 2006). Available from: http://www.fcc.gov/911/enhanced.

Fennell, D.A. (1996) 'A tourist space-time budget in the Shetland islands', Annals of Tourism Research, 23(4): 811–29.

Fennell, D.A. (2006) Tourism Ethics, Clevedon: Channel View Publications.

Fisher, P. and Dobson, J. (2003) 'Who knows where you are, and who should, in the era of mobile geography', Geography, 88(4): 331–37.

Forer, P. (1998) 'Geometric approaches to the nexus of time, space, and microprocess: Implementing a practical model for mundane socio-spatial systems', in M.J. Egenhofer and R.G. Golledge (eds) Spatial and Temporal Reasoning in Geographic Information Systems, Oxford: Oxford University Press, 171–90.

Forer, P. (2002) 'Tourist flows and dynamic geographies', in D.G. Simmons, and J. Fairweather (eds) Understanding the Tourism Host-Guest Encounter in New Zealand: Foundations for Adaptive Planning and Management Christchurch: EOS Ecology, 21–56.

Foroohar, R. (2003) 'The all-seeing eyes: New mobile phones can find almost anything, including you', Newsweek, 15 December.

Frew, E.A. and Shaw, R. (1999) 'The relationship between personality, gender, and tourism behavior', Tourism Management, 20: 193–202.

Freytag, T. (2002) 'Tourism in Heidelberg: Getting a picture of the city and its visitors', in K. Wöber (ed.) City Tourism 2002: Proceedings of European Cities Tourism's International Conference in Vienna, Austria. Vienna: Springer, 211–19.

Gali-Espelt, N. and Donaire-Benito, J.A. (2006) 'Visitors' behavior in heritage cities: The case of Girona', Journal of Travel Research, 44(4): 442–48.

Garbrecht, D. (1971) 'Pedestrian paths through a uniform environment', Town Planning Review, 41: 71–84.

Gärling, T., Brännäs, K., Garvill, J., Golledge, R.G. Gopal, S., Holm, E., and Lindberg, E. (1989) 'Household activity scheduling', in Transport Policy, Management and Technology Towards 2001: Selected proceedings of the Fifth World Conference on Transport Research vol. 4, Ventura, CA: Western Periodicals, 235–248.

Getting, I.A. (1993) 'The global positioning system', IEEE Spectrum, 30(12): 36–47.

Gladstone, D.L. (1998) 'Tourism urbanization in the United States', Urban Affairs Review, 34(1): 3–27.

Golledge, R.G., Klatzky, R.L., Loomis, J.M., Speigle, J., and Tietz, J. (1998) 'A geographical information system for a GPS based personal guidance system', International Journal of Geographical Information Science, 12(7): 727–49.

Golledge, R.G., Kwan, M.P., and Gärling, T. (1994) 'Computational process modeling of household travel decisions using a geographical information system', Proceedings of the Regional Science Association, 41: 169–204.

Golledge, R.G., Loomis, J.M., Klatzky, R.L., Flury, A., and Yang, X. (1991) 'Designing a personal guidance system to aid navigation without sight: Progress on the GIS component', International Journal of Geographical Information Systems, 5: 373–95.

Golledge, R.G. and Stimson, R.J. (1987) Analytical Behavioural Geography, London: Croom Helm.

Golledge, R.G. and Stimson, R.J. (1997) Spatial Behavior: A Geographic Perspective, New York and London: The Guilford Press.

Golledge, R. G. and Timmermans, H. (eds) (1988) Behavioural Modelling in Geography and Planning, London: Croom Helm.

Golob, T.F. and McNally, M.G. (1997) 'A model of activity participation and travel interactions between household heads', Transportation Research B, 31(3): 177–94.

Goodchild, M.F., Anselin, L., Appelbaum, R.P., and Herr-Harthorn, B. (2000) 'Towards spatially integrated social science', International Regional Science Review, 23(2): 139–59.

Goodchild, M.F. and Janelle, D.G. (1984) 'The city around the clock: Space-time patterns of urban ecological structure', Environment and Planning A, 16: 807–20.

Gottdiener, M. (1994) The New Urban Sociology, New York: McGraw-Hill.

Gottdiener, M. (2000) Life in the Air: Surviving the New Culture of Air Travel, Lanham: Rowman and Littlefield.

Gregory, D. (1989) 'Areal differentiation and post-modern human geography', in D. Gregory and R. Walford (eds) Horizons in Human Geography, Totowa, NJ: Barnes and Noble Books, 67–96.

Gregory, D. (2000) 'Time-geography', in R.J. Johnston, D. Gregory, G. Pratt, and M. Watts (eds) The Dictionary of Human Geography, 830–3. Oxford and Malden, MA: Blackwell.

Gren, M. (2001). 'Time-geography matters', in J. May and N. Thrift (eds) Timespace: Geographies of Temporality, London and New York: Routledge, 208–25.

Hägerstrand, T. (1953) Innovationsförloppet Ur Korologisk Synpunkt, Gleerup: Lund. [English translation 1967 Innovation diffusion as a spatial process by A. Pred. Chicago.]

Hägerstrand, T. (1970) 'What about people in regional science?', Papers of the Regional Science Association, 24(1): 7–21.

Haklay, M., O'sullivan, D., and Thurstain-Goodwin, M (2001) '"So go downtown": Simulating pedestrian movement in town centres', Environment and Planning B, 28(3): 343–59.

Hall, C. M. (2005) 'Reconsidering the geography of tourism and contemporary mobility', Geographical Research, 43(2): 125–139.

Hall, C.M. and Page, S.J. (1999) The Geography of Tourism and Recreation: Environment, Place and Space, London and New York: Routledge.

Hall, C.M. and Page, S.J. (2002 2nd edition) The Geography of Tourism and Recreation: Environment, Place and Space, Second Edition, London: Routledge.

Hall, P. (1970) 'A horizon of hotels', New Society, 15: 389–445.

Halpin, B. and Chan, T.K. (1998) 'Class careers as sequences: An optimal matching analysis of work-life histories', European Sociological Review, 14(2): 111–30.

Hanson, S. and Hanson, P. (1981) 'Travel-activity patterns of urban residents: Dimensions and relationships to socio-demographic characteristics', Economic Geography, 54: 332–47.

Hartmann, R. (1988) 'Combining field methods in tourism research', Annals of Tourism Research, 15: 88–105.

Helbing, D., Molnar, P., Farkas, I. J., and Bolay, K (2001) 'Self-organizing pedestrian movement', Environment and Planning B, 28(3): 361–83.

Hill, M. (1984) 'Stalking the urban pedestrian: A comparison of questionnaire and tracking methodologies for behavioral mapping in large-scale environments', Environment and Behavior, 16: 539–50.

Hovgesen, H.H., Bro, P., Tradisauskas, N., and Nielsen, T.S. (2008) 'Tracking visitors in public parks: Experiences with GPS in Denmark', in J. van Schaick and S. van der Spek (eds) Urbanism on Track: Application if Tracking Technologies in Urbanism, Amsterdam: IOS Press, 65–78.

Huff, J.O. and Hanson, S. (1986) 'Repetition and variability in urban travel', Geographical Analysis, 18(2): 97–114.

Huff, J.O. and Hanson, S. (1990) 'Measurement of habitual behavior: Examining systematic variability in repetitive travel. In P.M. Jones (ed.) Developments in Dynamic and Activity-Based Approaches to Travel Analysis, Aldershot: Gower, 229–49.

Hughes, H.L. (1997) 'Holidays and homosexual identity', Tourism Management, 18: 3–7.

Hughes, H.L. (1998) 'Sexuality, tourism and space: The case of gay visitors to Amsterdam', in D. Tyler, Y. Guerrier and M. Robertson (eds) Managing Tourism in Cities: Policy, Process and Practice, Chicester: John Wiley & Sons.

Huisman, O., P. Forer. (1998) 'Towards a geometric framework for modelling space-time opportunities and interaction potential', Paper presented at the International Geographical Union, Commission on Modeling Geographical Systems Meeting (IGU-CMGS), Lisbon, Portugal, 28–29 August.

Israel, M. and Hay, I. (2006) Research Ethics for Social Scientists: Between Ethical Conduct and Regulatory Compliance, Thousand Oaks, California: Sage.

Janelle, D.G. (2004) Time and the City: Synoptic Analysis of Space-time Activity Systems Executive Summary and Policy Proposal Venice, International Observatory on Sustainable Mobility in metropolitan areas.

Janelle, D.G. (2005) Synoptic Analysis of Space-time Activity Patterns. A position statement for the FHWA-sponsored Peer Exchange and CSISS Specialist Meeting––GPS and Time-Geography Applications for Activity Modeling and Microsimulation Center for Spatially Integrated Social Science, University of California, Santa Barbara 10–11 October.

Janelle, D.G., Goodchild, M. F., and Klinkenberg. K, (1988) 'Space-time diaries and travel characteristics for different levels of respondent aggregation', Environment and Planning A, 20(7): 891–906.

Jansen-Verbeke, M. (1991) 'The synergy between shopping and tourism: The Japanese experience', in W.F. Theobald (ed.) Global Tourism: The Next Decade, Oxford: Butterworth—Heinemann.

Jefferson A. and Lickorish, L. (1988) Marketing Tourism: A Practical Guide, Harlow, Essex: Longman.

Joh, C.H., Arentze, T.A., and Timmermans H.J.P. (2001) 'A position-sensitive sequence-alignment method illustrated for space-time activity-diary data', Environment and Planning A, 33(2): 313–38.

Joh, C.H., Arentze, T.A., Hofman, F., and Timmermans, H.J.P. (2002) 'Activity pattern similarity: A multidimensional sequence alignment method', Transportation Research Part B: Methodological, 36(5): 385–403.

Joh, C.H., Arentze, T.A., and Timmermans H.J.P. (2005) 'A utility-based analysis of activity time allocation decisions underlying segmented daily activity-travel patterns', Environment and Planning A, 37(1): 105–25.

Jules-Rosette, B. (1994) 'Black Paris: Touristic simulations', Annals of Tourism Research, 21: 679–700.

Kaplan, E.D. (1996) Understanding GPS, Principles and Applications, Norwood MA: Artech House.

Kemperman, A.D.A.M., Joh, C.H., and Timmermans H.J.P. (2004). 'Comparing first time and repeat visitors activity patterns' Tourism Analysis 8: 159–64.

Keul, A. and Küheberger, A. (1997) 'Tracking the Salzburg tourist', Annals of Tourism Research, 24(4): 1008–12.

Kitamura, R. Nishii, K. and Goulias, K. (1988) 'Trip chaining behavior by central city commuters: A causal analysis of time-space constraints', Paper presented at the Oxford Conference on Travel and Transportation.

Kitamura, R., Nishii, K., and Goulias, K. (1990) 'Trip chaining behavior by central city commuters: A causal analysis of time-space constraints', in P. Jones (ed.) Developments in Dynamic and Activity Based Approaches to Travel Analysis, Avebury: Aldershot, 145–70.

Knaap, W. G. M. van der (1999) 'Research report: GIS-oriented analysis of tourist time-space patterns to support sustainable tourism development', Tourism Geographies, 1, (1): 56–69.

Knafou, R. and Stock, M. (2003)'Tourisme', in J. Lévy, and M. Lussault (eds) Dictionnaire de la Géographie et de L'espace des Sociétés, Paris: Belin.

Knox, P.L. (1994) Urbanization: An Introduction to Urban Geography, Englewood Cliffs, NJ: Prentice Hall.

Kondo, K. and Kitamura, R. (1987) 'Time-space constraints and the formation of trip chains', Regional Science and Economics, 17(1): 49–65.

Kroes, E. (1990) Stated preference onderzoek: Veranderingen in de vertrektijd keuze onder invloed van congestie, The Hague: HCG.

Kumar, A. and Levinson, D.M. (1995) 'Chained trips in montgomery county, Maryland', ITE (Institute of Traffic Engineers) Journal, 27–32.

Kwan, M.P. (1998) 'Space-time and integral measures of individual accessibility: A comparative analysis using a point-based framework', Geographical Analysis, 30(3): 191–216.

Kwan, M.P. (1999a) 'Gender, the home-work link, and space-time patterns of non-employment activities', Economic Geography, 75(4): 370–394.

Kwan, M.P. (1999b) 'Gender and individual access to urban opportunities: A study using space-time measures', The Professional Geographer, 51(2): 210–27.

Kwan, M.P. (2000) 'Interactive geovisualization of activity-travel patterns using three-dimensional geographical information systems: A methodological exploration with a large data set', Transportation Research C, 8:185 203.

Kwan, M.P. (2002a) 'Time, information technologies, and the geographies of everyday life', Urban Geography, 23(5): 471–82.

Kwan, M.P. (2002b) 'Feminist visualisation: Re-envisioning GIS as a method in feminist geographic research', Annals of the Association of American Geographers, 92(4): 645–61.

Kwan, M.P. (2004) 'GIS methods in time-geographic research: Geocomputation and geovisualization of human activity patterns', Geografiska Annaler B, 86(4): 267–80.

Kwan, M.P. and Lee, J. (2004) 'Geovisualization of human activity patterns using 3-D GIS: A time geographic approach', in N.F. Goodchild and D.G. Janelle (eds) Spatially Integrated Social Science, New York: Oxford University Press, 48–66.

Kwan, M.P. and Weber, J. (2003) 'Individual accessibility revisited: Implications for geographical analysis in the twenty-first century', Geographical Analysis, 35(4): 341–53.

Law, C.M. (1996) Tourism in Major Cities, London: Routledge.

Le Faucheur, A., Abraham, P., Jaquinandi, V., Bouyé, P., Saumet, J.L., and Noury-Desvaux, B. (2008) 'Measurement of walking distance and speed in patients with peripheral arterial disease', Circulation, 19: 897–904.

Lembke, J. (2003) 'EU critical infrastructure and security policy: Capabilities strategies and vulnerabilities', in F. H. Columbus (ed.) European Economic and Political Issues, New York: Nova Publishers, 49–79.

Lenntorp, B. (1976) 'Paths in space-time environment: A time geographic study of possibilities of individuals', Lund Studies in Geography Series B, Human Geography 44.

Lenntorp, B. (1999) 'Time-geography—at the end of its beginning', GeoJournal, 48(3): 155–8.

Levy, S. (2004) 'A Future with Nowhere to Hide?', Newsweek, 7 June.

Lew A.A. (1986) Guidebook Singapore: The Spatial Organization of Urban Tourist Attractions, Ph.D. Dissertation. University of Oregon.

Lew, A.A. (1987) 'The english-speaking tourist and the attractions of Singapore', Singapore Journal of Tropical Geography, 8: 44–59.

Lew, A.A. and McKercher, B. (2004) Travel geometry: macro and micro scales considerations. Paper presented at the Pre-Congress Meeting of the International Geographic Union's Commission on Tourism, Leisure and Global Change. Loch Lomond, Scotland, 13th–15th August, 2004.

Lew, A.A. and McKercher, B. (2006) 'Modeling tourist movements: A local destination analysis', Annals of Tourism Research, 33(2): 402–23.

Light, D. and Prentice, R.C. (1994) 'Who consumes the heritage product ?: Implications for European heritage tourism', in G.J. Ashworth and P.J. Larkham (eds) Building A New Heritage: Tourism, Culture and Identity in the New Europe, London and New York: Routledge.

Longley, P.A. (2002) 'Geographical information systems: Will developments in urban remote sensing and GIS lead to a 'better' urban geography?', Progress in Human Geography, 26:231–9

Ludden, B., A. Pickford, J. Medland, H. Johnson, F. Bandon, L. E. Axelson, K. Viddal-Ervik, B. Dorgelo, E. Boroski, and Malenstein, J. (2002) Report on Implementation Issues Related to Access to Location Information by Emergency Services (E112) in the European Union. A Report Prepared for the Coordination Group on Access to Location Information for Emergency Services.

Lyon, D. (2001) Surveillance Society: Monitoring everyday life, Buckingham: Open University Press.

Maeda, Y., Eiichi, T., Hideo, M., Takashi, K., and Ikuo, I. (2002) 'Evaluation of a GPS-based guidance system for visually impaired pedestrians', in Proceedings of the Technology and Persons with Disabilities Conference 2002.

Mannering, F. (1989) 'Position analysis of commuter flexibility in changing route and departure times', Transportation Research B, 23: 53–60.

Mateos, P. and Fisher, P. (2006) 'Spatiotemporal accuracy in mobile phone location: Assessing the new cellular geography', in J. Drummond, R. Billen, E. João, and D. Forrest (eds) *Dynamic & Mobile GIS: Investigating Change in Space and Time,* London: Taylor & Francis.

McKercher, B. and Lau, G. (2008) 'Movement patterns of tourists within a dDestination', Tourism Geographies, 10(3): 355–74.

Meng, L., Zipf, A., and Reichenbacher, T. (eds) (2005) Map-based Mobile Services: Theories, Methods and Implementations, Berlin: Springer.

Mertens, D.M. and Ginsberg, P.E. (2008) The Handbook of Social Research Ethics, Thousand Oaks, California: Sage.

Merton, R.K. (1968) Social Theory and Social Structure, New York: Free Press.

Miller, H.J. (1991) 'Modeling accessibility using space-time prism concepts within geographical information systems', International Journal of Geographical Information Systems, 5: 287–301.

Miller, H.J. (2003) 'What about people in geographic information science?', Computers, environment and urban systems, 27(5): 447–53.

Miller, H.J. (2005) 'A measurement theory for time geography', Geographical Analysis, 37(1): 17–45.

Miskelly, F. (2004) 'A novel system of electronic tagging in patients with dementia and wandering', Age and Ageing, 33(3): 304–6.

Miskelly, F. (2005). 'Electronic tracking of patients with dementia and wandering using mobile phone technology', Age and Ageing, 34(5): 497–9.

Modsching, M., Kramer, R., Ten Hagen, K., and Gretzel, Y. (2008) 'Using location-based tracking data to analyze the movements of city tourists', Information Technology & Tourism, 10(1): 31–42.

Montanari, A. and Muscará, C. (1995) 'Evaluating tourist flows in historic cities: The case of Venice', Tidschrift voor Economische en Sociale Geographie, 86(1): 80–7.

Moore, A. (1985) 'Rosanzerusu is Los Angeles: An anthropological inquiry of Japanese tourists', Annals of Tourism Research, 12: 619–43.

Mullins, P. (1991) 'Tourism urbanization', International Journal of Urban and Regional Research, 15(3): 326–41.

Murphy, P.E. (1992) 'Urban tourism and visitor behavior', American Behavioral Scientist, 36: 200–11.

Murphy, P.E. Oppermann, M. (1997) 'First-time and repeat visitors to New Zealand', Tourism Management, 18: 177–81.

Murphy, P.E. and Rosenblood, L. (1974) 'Tourism: An exercise in spatial search', Canadian Geographer, 18: 201–10.

Needleman S.B. and Wunsch, C.D. (1970) 'A general method applicable to the search for similarities in the amino acid sequence of two proteins', Journal of Molecular Biology, 48: 443–53.

Oppermann, M. (1997) 'First-time and repeat visitors to New Zealand', Tourism Management, 18(3): 177–81.

Page, S.J. and Hall, M.C. (2003) Managing Urban Tourism, Harlow: Prentice Hall.

Palm, R. and Pred, A. (1974) 'A time-geographic perspective on problems of inequality for woman', Working Paper No. 26 Institute of Urban and Regional Development, University of California, Berkley.

Parks, D.N. and Thrift, N. (1980) Times, Spaces and Places: A Chronogeographic Perspective, Chichester, UK: John Wiley & Sons.

Parkinson, B.W. (1994) 'GPS eyewitness: The early years', GPS World, 5(9): 32–45.

Pas, E.I. (1983) 'A flexible and integrated methodology for analytical classification of daily travel-activity behavior', Transportation Science, 17(3): 405–29.

Pas E.I. and Koppelmann, F.S. (1986) 'An examination of day to day variability in individuals' urban travel behavior', Transportation, 13(2): 183–200.

Pearce, D.G. (1987) Tourism Today: a geographical analysis. Burnt Mill: Longman.

Pearce, D.G. (1988) 'Tourist time-budgets', Annals of Tourism Research, 15: 106–21.

Pearce, D.G. (1995, 2nd edition) Tourism Today: A Geographical Analysis, Burnt Mill: Longman.

Pendyala, R.M., Kitamura, R., and Reddy, D.V.G.P. (1998) 'Application of an activity-based travel demand model incorporating a rule based algorithm', Environment and Planning B, 25: 753–72.

Phillips, M.L., Hall, T.A., Esmen, N.A., Lynch, R., and Johnson, D.L. (2001) 'Use of global positioning system technology to track subject's location during environmental exposure sampling', Journal of Exposure Analysis and Environmental Epidemiology, 11: 207–15.

Pizam, A. and Jeong, G.H. (1996) 'Cross-cultural tourist behaviour: Perceptions of Korean tour-guides', Tourism Management, 17: 277–86.

Pizam, A. and Sussmann, S. (1995) 'Does nationality affect tourist behavior', Annals of Tourism Research, 22: 901–17.

Plog, S.C. (1973) 'Why destination areas rise and fall in popularity', Cornell Hotel and Restaurant Administration Quarterly, 14(4): 13–16.

Plog, S.C. (1987) 'Understanding psychographies in tourism research', in J. R. Brent-Ritchie and C.R. Goeldner (eds), Travel, Tourism and Hospitality Research: A Handbook for Managers and Researchers, New York: John Wiley.

Posner, R.A. (1978) 'The Right of Privacy' Georgia Law Review, 12: 393–422.

Post, R. (1989) 'The social foundations of privacy: Community and self in the common law tort', California Law Review, 77: 957–76.

Prentice, R. (1993) Tourism and heritage attractions, London and New York: Routledge.

Prideaux, B. (2000) 'The role of the transport system in destination development', Tourism Management, 21: 53–63.

Prosser, W.L. (1960) 'Privacy', California Law Review, 48: 383–423.

Quiroga, C.A. and Bullock, D. (1998) 'Travel time studies with global positioning and geographic information systems: An integrated methodology', Transportation Research C, 6(1): 101–27.

Rachels, J. (1975) 'Why privacy is important', Philosophy and Public Affairs, 4: 323–33.

Ratti, C., Pulselli, R.M., Williams, S., and Frenchman, D. (2006) 'Mobile landscapes: Using location data from cell-phones for urban analysis', Environment and Planning B, 33(3): 727–48.

Ratti, C., Sevtsuk, A., Huang, S., and Pailer, R. (2005) 'Mobile landscapes: Graz in real time', Proceedings of the 3rd Symposium on LBS & TeleCartography, 28–30 November Vienna

Raubal, M., Miller, H.J., and Bridwell, S. (2004) 'User-centered time geography for location-based services', Geografiska Annaler B, 86(4): 245–65.

Reades, J. (2008) 'People, places & privacy', International Workshop Social Positioning Method (SPM) 2008. Tartu, Estonia, 10–14 March, 2008. (Can be accessed at: http://www.reades.com/ privacy/).

Reades, J., Calabrese, F., Sevstuk, A., and Ratti, C. (2007) 'Cellular census: Explorations in urban data collection', Pervasive Computing, 6(3): 30–8.

Recker, W.W., McNally, M.G., and Root, G.S. (1983) 'Application of pattern recognition theory to activity pattern analysis', in S. Carpenter, and P. Jones (eds) Recent Advances in Travel Demand Analysis, Aldershot: Gower, 434–49.

Recker, W.W., McNally, M.G., and Root, G.S. (1987) 'An empirical analysis of urban activity patterns', Geographical Analysis, 19(2): 166–81.

Renenger, A. (2002) 'Satellite tracking and the right to privacy', Hastings Law Journal, 53: 549–65.

Richardson, S.L. and Crompton, J. (1988) 'Vacation patterns of French and English Canadians', Annals of Tourism Research, 15: 430–48.

Richtel, M. (2005) 'Enlisting cellphone signals to fight road gridlock', The New York Times, 11 November.

Russo, A.P. (2001) 'The vicious circle of tourism development in heritage cities', Annals of Tourism Research, 29(1): 165–82.

Saitou, N. and Nei, M. (1987) 'The neighbor-joining method: A new method for reconstructing phylogenetic trees', Molecular Biology and Evolution, 4(4): 406–25.

Sankoff, D. and Kruskal, J. (1983) Time Warps, String Edits, and Macromolecules: The Theory and Practice of Sequence Comparison, Reading, MA: Addison-Wesley.

Schaick, J. van and Spek, S. van der (Eds) (2008). Urbanism on Track. Delft University Press.

Schlich, R. and Axhausen, K.W. (2003) 'Habitual travel behaviour: Evidence from a six-week travel diary', Transportation, 30(1): 13–36.

Schilling, A., Coors, V., and Laakso, K. (2005) 'Dynamic 3D maps for mobile tourism applications. in L. Meng, A. Zipf and T. Reichenbacher (eds) Map-based Mobile Services: Theories, Methods and Implementations, New York: Springer Geosciences, 233–44.

Scraton, S. and Watson, B. (1998) 'Gendered cities: Women and public leisure space in the postmodern city', Leisure Studies, 17: 122–37.

Shachar, A. (1995) 'Metropolitan areas: Economic globalisation and urban tourism', in A. Montanari and A.M. Williams (eds) European Tourism: Regions, Spaces and Restructuring, Chichester: John Wiley & Sons, 151–160.

Shachar, A. and Shoval, N. (1999) 'Tourism in Jerusalem: A place to pray', in D. Judd and S.S. Fainstein (eds) The Tourist City, New Haven: Yale University Press, 198–211.

Shaw, G., Agarwal, S., and Bull, P. (2000) 'Tourism consumption and tourism behavior: A British perspective', Tourism Geographies, 2: 264–89.

Shaw, G. and Williams, A.M. (2002) Critical Issues in Tourism: A Geographical Perspective, Second Edition, Oxford: Blackwell.

Shaw, G., Williams, A.M., and Greenwood, J. (1990) Visitor Patterns and Visitor Behavior in Plymouth, Exeter: Tourism Research Group, University of Exeter.

Sheridan, B, (2009) 'A Trillion Points of Data', Newsweek, March 9, 32–35.

Shoval, N. (2001) Segmented and Overlapping Tourist Spaces: Jerusalem and Tel-Aviv as Case Studies, Ph.D. dissertation, The Hebrew University of Jerusalem.

Shoval, N. (2002) 'Spatial activity of tourists in cities: What are the underlying factors?', in K.W. Wöber (ed.) City Tourism 2002, Vienna and New York: Springer, 18–33.

Shoval, N. (2006) 'The geography of hotels in cities: An Empirical validation of a forgotten theory, Tourism Geographies, 8(1): 56–75.

Shoval, N. (2007) 'Sensing human society', Environment and Planning B: Planning and Design, 34: 191–5.

Shoval, N. (2008) 'Tracking technologies and urban analysis', Cities, 25(1): 21–8.

Shoval, N., Auslander, G.K., Freytag, T., Landau, R., Oswald, F., Seidl, U., Wahl, H.W., Werner, S., and Heinik, J. (2008) 'The use of advanced tracking technologies for the analysis of mobility in Alzheimer's disease and related cognitive diseases', BMC Geriatrics, 8(7).

Shoval, N. and Isaacson, M. (2006) 'The application of tracking technologies to the study of pedestrian spatial behaviour', The Professional Geographer, 58(2): 172–83.

Shoval, N. and Isaacson, M. (2007a) 'Tracking tourist in the digital age', Annals of Tourism Research, 34(2): 141–59.

Shoval, N. and Isaacson, M. (2007b) 'Sequence alignment as a method for human activity analysis', Annals of the Association of American Geographers, 97(2): 282–97.

Shoval, N. and Raveh, A. (2004) 'The categorization of tourist attractions: The modeling of tourist cities based on a new method of multivariate analysis', Tourism Management, 25(6): 741–50.

Sieberg D. (2006) 'Is RFID tracking you?', CNN.com http://edition.cnn.com /2006/TECH/07/10/rfid/.

Skov-Petersen, H. (2005) 'Feeding the agents: Collecting parameters for agent-based models' CUPUM,. London 29/6–1/7 2005.

Smith, V.L. (ed.) (1977) Hosts and Guests: The Anthropology of Tourism, Philadelphia: University of Pennsylvania Press.

Spek, S. van der (2008A) 'Tracking Technologies: An Overview', in J. van Schaick and S. van der Spek (eds) Urbanism on Track: Application if Tracking Technologies in Urbanism, Amsterdam: IOS Press. 25–33.

Spek, S. van der (2008B) 'Spatial Metro: Tracking Pedestrians in Historic City Centres', in J. van Schaick and S. van der Spek (eds) Urbanism on Track: Application if Tracking Technologies in Urbanism, Amsterdam: IOS Press. 79–102.

Stovel, K. and Bolan, M. (2004) 'Residential trajectories: Using optimal alignment to reveal the structure of residential mobility', Sociological Methods Research, 32(4): 559–98.

Szalai, A. (ed.) (1972) The Use of Time: Daily Activities of Urban and Suburban Populations in Twelve Countries, The Hague and Paris: Mouton.

Ten Hagen, K., Modsching, M., and Kramer, R. (2005) A Location Aware Mobile Tourist Guide Selecting and Interpreting Sights and Services by Context Matching. Paper presented at the 2nd Annual International Conference on Mobile and Ubiquitous Systems: Networking and Services, 17–21 July, San Diego, CA.

Terrier, P. and Schutz, Y. (2005). 'How useful is a satellite positioning system (GPS) to track gait parameters? A review', Journal of NeuroEngineering and Rehabilitation, 2(28).

The White House. Statement by the Press Secretary September 2007. http://www. whitehouse.gov/news/releases/2007/09/20070918-2.html.

The World Bank (2006) World Bank Development Indicators 2006, Washington, DC: The World Bank.

Thomson, J. J. (1975) 'The Right to Privacy', Philosophy and Public Affairs 4(4): 295–314.

Thornton, P.R., Williams, A.M., and Shaw, G. (1997) 'Revisiting time-space diaries: An exploratory case study of tourist behaviour in Cornwall, England', Environment and Planning A, 29(10): 1847–67.

Thrift, N. (2002) 'The future of geography', Geoforum, 33(3): 291–98.

Thurstain-Goodwin, M. (2003) 'Data surfaces for a new policy geography', in P.A. Longley and M. Batty (eds) Advanced Spatial Analysis: The CASA book of GIS, Redlands, California: ESRI Press, 145–170.

Timmermans, H.J.P., Arentze, T., and Chang-Hyeon J. (2002) 'Analyzing space-time behavior: New approaches to old problems', Progress in Human Geography, 26(2): 175–90.

Timmermans, H.J.P. and van der Waerden, P.J.H.J. (1993) 'Modelling sequential choice processes: The case of two-stop trip-chaining', Environment and Planning A, 24: 1483–90.

Towner, J. (1996) An Historical Geography of Recreation and Tourism in the Western World 1540–1940, Chichester: Wiley.

Veal, A.J. (1992) Research Methods for Leisure and Tourism: A Practical Guide, London: Longman.

Warren, S. and Brandeis, L. (1890) 'The right to privacy', Harvard Law Review, 4(1): 193–220.

Waterman, M. (1995) Introduction to Computational Biology, London: Chapman and Hall.

White, R. and Engelen, G. (2000) 'High-resolution integrated modeling of the spatial dynamics of urban and regional systems', Computers, Environment and Urban Systems, 24: 383–400.

Wilson, C. (1998) 'Activity pattern analysis by means of sequence-alignment methods', Environment and Planning A, 30(6): 1017–38.

Wilson, C. (2001) 'Activity patterns of Canadian women: Application of ClustalG sequence alignment software', Transportation Research Record, 1777: 55–67.

Wilson, C. (2006) 'Reliability of sequence Alignment Analysis of social processes: Monte Carlo Tests of ClustalG software', Environment and Planning A, 38(1): 187–204.

Wilson, C. (2008) 'Activity patterns in space and time: Calculating representative Hagerstrand trajectories', Transportation, 35: 485–99.

Wilson, C., Harvey, A. and Thompson, J. (1999) 'ClustalG: Software for analysis of activities and sequential events', Paper presented at the Workshop on Longitudinal Research in Social Sciences: A Canadian Focus, Windermere Manor, London, Ontario, Canada.

Wolf, J., Guensler, R., and Bachman, W. (2001) 'Elimination of the travel diary: Experiment to derive trip purpose from GPS travel data', Paper presented at the 80th Annual Meeting of the Transportation Research Board, 7–11 January, Washington, D.C.

Xia, J., Arrowsmith, C., Jackson, M., and Cartwright, W. (2008) 'The wayfinding process relationships between decision-making and landmark utility', Tourism Management, 29: 445–57.

Xu, G. (2007) GPS: Theory, Algorithms and Applications, New York: Springer

Yamamoto, T. and Kitamura, R. (1999) 'An analysis of time allocation to in-home and out of-home discretionary activities across working days and non-working days', Transportation, 26: 211–30.

Yan, W. and Forsyth, D. (2005) 'Learning the behavior of users in a public space through video tracking", in Proceedings of IEEE Workshop on Applications of Computer Vision (WACV) 2005, Breckenridge Colorado.

Yokeno, N. (1968) La localisation de l'industrie touristique: Application de l'analyse de Thunen-Weber, Cahiers du Tourisme, C-9, C. H. E. T., Aix-en-Provence.

Zhao, Y. (1997) Vehicle Location and Navigation Systems, Norwood MA: Artech House.

Zillinger, M. (2007) 'Tourist routes: A time-geographical approach on German car-tourists in Sweden', Tourism Geographies, 9(1): 64–83.

Zito, R., D'este, G., and Taylor, M.A.P. (1995) 'Global positioning in the time domain: How useful a tool for intelligent vehicle-highway systems?', Transportation Research C, 3(4): 193–209.

Index